Systems Biology of Apoptosis

Inna N. Lavrik
Editor

Systems Biology of Apoptosis

Springer

Editor
Inna N. Lavrik
Department of Translational Inflammation Research
Institute of Experimental Internal Medicine
Otto von Guericke University
Magdeburg, Germany

ISBN 978-1-4614-4008-6 ISBN 978-1-4614-4009-3 (eBook)
DOI 10.1007/978-1-4614-4009-3
Springer New York Heidelberg Dordrecht London

Library of Congress Control Number: 2012943946

Printed on acid-free paper

Springer is part of Springer Science+Business Media (www.springer.com)

Preface

Programmed cell death is a fascinating process common to all multicellular organisms. Programmed cell death results in the elimination of cells via a complex but a highly defined programme. Defects in the regulation of programmed cell death are associated with serious diseases such as cancer, autoimmunity, AIDS, and neurodegeneration.

Apoptosis has been the best studied type of programmed cell death so far. Cells that undergo apoptosis are characterized by chromatin condensation, nuclear fragmentation, membrane blebbing, cell shrinkage, and formation of apoptotic bodies.

The central role in apoptosis execution belongs to cysteine-specific aspartate proteases (caspases). Caspases are enzymes that orchestrate apoptosis via cleavage of cellular substrates.

There are two major pathways of apoptosis: intrinsic and extrinsic. The intrinsic pathway is triggered via chemotherapeutic drugs, irradiation, and growth factor withdrawal. These stimuli lead to mitochondrial outer membrane permeabilization (MOMP), which results in cytochrome *C* release and caspase activation. In the extrinsic apoptotic pathway, the caspase cascade is triggered by signals emanating from the cell-surface death receptors (DR) triggered by death ligands (DL) (TNF, CD95L/FasL, TRAIL). The DR stimulation results in the formation of the death-inducing signaling complex (DISC) and subsequent caspase activation.

Despite the fact that signaling pathways of apoptosis have been described with an impressive level of detail, the understanding of apoptosis regulation in quantitative terms has been missing until recently. There were many unclear points: when does a cell decide that it has to die, what are the rate-limiting steps in apoptosis, is there a point of no return, how can cell death be accelerated or blocked, and many others. From another side the years of apoptosis research resulted in a profound understanding of how signaling in apoptosis occurs. All major apoptotic complexes have been identified from the DISC to the apoptosome, including the death receptors and adaptors and the most important enzymes and their inhibitors. Therefore, apoptosis was an ideal system to go into quantitative studies using the emerging field of systems biology.

Systems biology combines mathematical modeling with experimental approaches in a closed loop cycle (Fig. 1). On the modeling side there are a number of mathematical formalisms, e.g., Ordinary Differential Equations (ODEs), Boolean models, etc., that allow to address different biological questions. Experimental work for systems biology of apoptosis involves the generation of quantitative data using different apoptotic assays.

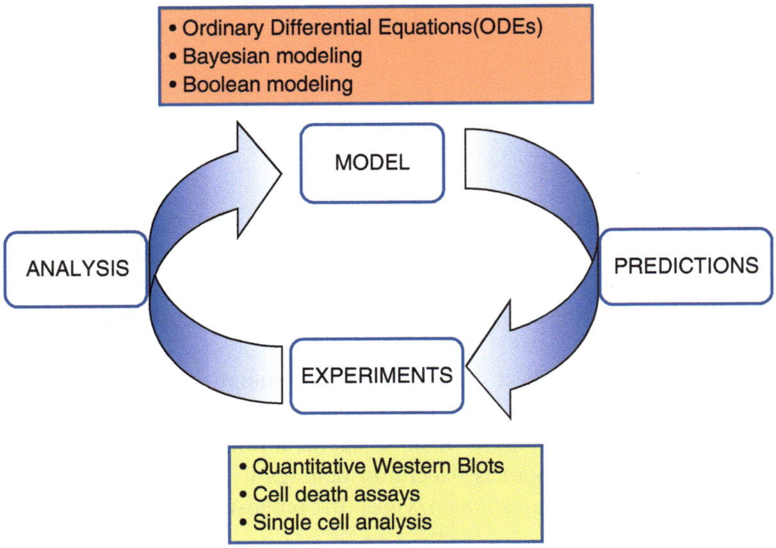

Fig. 1 Systems biology of apoptosis. Schematic view of systems biology of apoptosis

The development of this field in the recent years is fascinating. Studies of apoptosis using systems biology have provided novel insights into the quantitative regulation of cell death. In this book we describe contemporary systems biology studies devoted to cell death signaling both from experimental and modeling sides and focus on the question how systems biology helps to understand life/death decisions made in the cell and how to develop new approaches to rational treatment strategies.

Chapter 1 starts with an overview of the major types of mathematical modeling used in apoptosis and cell death. A simple minimalistic model of CD95/Fas-induced apoptosis is designed to introduce the most commonly used mathematical formalism, ordinary differential equations (ODEs). Besides ODEs, other modeling approaches are discussed in depth as well.

In Chap. 2 we focus on the biology of the extrinsic apoptotic pathway and its modeling by Ordinary Differential Equations (ODEs). We discuss new insights in the extrinsic death signaling which have been obtained using modeling.

Chapter 3 is devoted to model reduction approaches and uses the extrinsic apoptotic signaling as an example. This chapter provides a beautiful example of how complex biological signaling can be simplified using mathematical modeling and how a simplified model can provide new insights in complex biological questions. The first three chapters provide a major insight into the modeling of extrinsic pathways.

Chapter 4 covers the molecular mechanisms of the mitochondrial apoptotic pathway and the major models describing this pathway. An emerging question in the field how bioenergetics influence the cell death pathway is also addressed in detail.

Chapter 5 further addresses the molecular mechanisms of extrinsic and intrinsic apoptosis in the context of modeling hepatocytes. Notably, an enormous progress has been recently made in modeling the signaling pathways in the liver and, in particular, cell death in the liver. This work is essential to define new therapeutic strategies for liver regeneration and liver disease.

Chapter 6 explores other forms of cell death, e.g., necrosis, autophagy, their cross talk with apoptosis, as well as the way to model cross talk between different cell types using Boolean modeling.

Chapter 7 describes a single cell analysis. Single cell analysis is compared to bulk approaches and the importance to follow a single cell rather than a cell population is discussed.

Chapter 8 discusses a systems-level understanding of cytokine–cytokine cross talk, namely how the cross talk between different cytokine pathways could be modeled on intracellular and extracellular levels. The importance of this cross talk for development and disease is also highlighted.

Chapter 9 deals with an important question in the field: the importance of searching for new components of cell death networks using different screening techniques. The unraveling of new components versus the investigation of dynamic models, which include all known components of the network, is a highly discussed question.

Taken together, the different chapters of the book describe in detail the remarkable progress which was made in recent years in systems biology of apoptosis and show new challenges in this field that can provide even more exciting insights into cell death regulation.

Contents

Contributors

Christoph Borner Institute of Molecular Medicine and Cell Research, Albert Ludwigs University Freiburg, Freiburg, Germany

Spemann Graduate School of Biology and Medicine (SGBM), Albert Ludwigs University Freiburg, Freiburg, Germany

Bioss – Centre for Biological Signaling Studies, Albert Ludwigs University Freiburg, Freiburg, Germany

Michael Boutros German Cancer Research Center, Division of Signaling and Functional Genomics, Heidelberg University, Department of Cell and Molecular Biology, Medical Faculty Mannheim, Heidelberg, Germany

Laurence Calzone Institut Curie, Paris, France

Ecole des Mines ParisTech, Paris, France

INSERM U900, Paris, France

Christopher D. Deppmann Department of Biology, University of Virginia, Charlottesville, VA, USA

Michael Ederer Institute for System Dynamics, University of Stuttgart, Stuttgart, Germany

Christina Falschlehner Division of Signaling and Functional Genomics, German Cancer Research Center (DKFZ), University of Heidelberg, Heidelberg, Germany

Nicolai Fricker Institute of Bioinformatics and Systems Biology, Helmholtz Zentrum München, Neuherberg, Germany

Division of Immunogenetics, German Cancer Research Center (DKFZ), Heidelberg, Germany

Bioquant, Heidelberg, Germany

Heinrich J. Huber Department of Physiology and Medical Physics, Centre for Systems Medicine, Royal College of Surgeons in Ireland, Dublin, Ireland

Kevin A. Janes Department of Biomedical Engineering, University of Virginia, Charlottesville, VA, USA

Stefan Kallenberger Division of Theoretical Bioinformatics, DKFZ and BioQuant, Heidelberg, Germany

Inna N. Lavrik Department of Translational Inflammation Research, Institute of Experimental Internal Medicine, Otto von Guericke University, Magdeburg, Germany

Stefan Legewie Institute of Molecular Biology, Mainz, Germany

Irmgard Merfort Department of Pharmaceutical Biology and Biotechnology, Albert Ludwigs University Freiburg, Freiburg, Germany

Carla O'Connor Centre for Human-Systems Medicine, Department of Physiology and Medical Physics, Royal College of Surgeons in Ireland, Dublin, Ireland

Jochen H.M. Prehn Department of Physiology and Medical Physics, Centre for Systems Medicine, Royal College of Surgeons in Ireland, Dublin 2, Ireland

Dennis Rickert Institute of Bioinformatics and Systems Biology, Helmholtz Zentrum München, Neuherberg, Germany

Rebekka Schlatter Institute for System Dynamics, University of Stuttgart, Stuttgart, Germany

Kolja Schleich Division of Immunogenetics, German Cancer Research Center (DKFZ), Heidelberg, Germany

Bioquant, Heidelberg, Germany

Kathrin Schmich Department of Pharmaceutical Biology and Biotechnology, Albert Ludwigs University Freiburg, Freiburg, Germany

Fabian J. Theis Institute of Bioinformatics and Systems Biology, Helmholtz Zentrum München, Neuherberg, Germany

Boris Zhivotovsky Institute of Environmental Medicine, Stockholm, Sweden

Andrei Zinovyev Institut Curie, Paris, France

Ecole des Mines ParisTech, Paris, France

INSERM U900, Paris, France

Chapter 1
Modeling Formalisms in Systems Biology of Apoptosis

Stefan Kallenberger and Stefan Legewie

Abstract Apoptosis is a form of cellular suicide central to various aspects in biology including tissue homeostasis, embryonic development, carcinogenesis, and neurodegenerative disorders. Quantitative modeling approaches provided valuable insights into the digital and irreversible nature of apoptosis initiation. In this chapter, we summarize the mathematical formalisms used in systems biology of apoptosis. In addition, we give an overview of apoptosis-related research questions that can be addressed by modeling. Moreover, we review top-down and bottom-up modeling approaches applied to apoptosis, and particularly focus on ordinary differential equation (ODE) modeling. Basic concepts such as bistability and sensitivity analysis are introduced, and a review of apoptosis-related ODE models is provided. We describe bistability, temporal switching, crosstalk between death and survival, and also discuss approaches to model cell-to-cell variability.

1.1 Why Modeling Apoptosis?

Apoptosis is a phenomenologically easily observable process. However, understanding its mechanistic basis is challenging owing to complex interactions of a large number of signaling proteins and emergent behavior at the systems level. After applying a sufficiently strong death-inducing stimulus to a population of cells, irreversible signaling events are initiated leading to the characteristic appearance of an apoptotic cell: Membrane blebbing proceeds, the cell shrinks, and organelles disintegrate. Apoptosis occurs for extrinsic stimuli on a timescale of hours and for intrinsic stimuli of days, and is accessible to several experimental techniques

---

S. Kallenberger
Division of Theoretical Bioinformatics, DKFZ and BioQuant, Heidelberg, Germany

S. Legewie (✉)
Institute of Molecular Biology, Mainz, Germany
e-mail: s.legewie@imb-mainz.de

I.N. Lavrik (ed.), *Systems Biology of Apoptosis*, DOI 10.1007/978-1-4614-4009-3_1,

allowing for the acquisition of quantitative data. The classical techniques of Western blotting and immunoprecipitation enable coincidental acquisition of coarsely time-resolved population data for proteins and their intermediate processing stages. Fluorescence-based flow cytometry techniques allow measuring the protein concentrations at the single-cell level. A major disadvantage of flow cytometry is the inability of tracking time-dependent behavior of individual cells. This problem is overcome by fluorescence-based microscopic methods that were developed to obtain quantitative data of single cells with high temporal resolution: The activity of caspases can be monitored with FRET reporters or smart probes that harbor caspase cleavage sites. Moreover, the mitochondrial pathway of apoptosis can be monitored by measuring Bax translocation, outer membrane permeabilization, and Smac release. The wide range of available experimental techniques and the detailed knowledge about molecular events render apoptosis a system suitable for modeling analyses. Apoptosis induced by death ligands is one of the few cell fate decisions known to proceed by purely posttranscriptional mechanisms, thus further simplifying the formulation of mathematical models.

Even though individual steps of the apoptotic signal transduction cascades are well understood, we lack insights into the system properties and the dynamics of the death decision. Questions to be addressed in apoptosis by systems biology approaches include:

1. How do cells ensure that apoptosis robustly occurs in all-or-none manner? What is the "point-of-no-return" representing irreversibility in apoptosis? Which signaling motifs are responsible for such digital and history-dependent behavior? As detailed below, mechanisms proposed using kinetic modeling include bistability due to positive feedback and sigmoidal responses arising from competitive inhibition.
2. How is specificity in the apoptosis vs. survival responses ensured? A topic of particular interest for apoptosis modeling is that apoptotic stimuli trigger survival or death signaling depending on initial conditions and the stimulus strength. At least in some cases, the inhibitory crosstalk between survival and cell death signaling pathways appears to be mutually exclusive at the single-cell level (Nair et al. 2004), implying that death and survival represent different attractor states for the cell. Modeling can be employed to identify critical nodes of signaling crosstalk that tip the balance between cell death and survival. Furthermore, the interlocked regulation of cell cycle is a topic followed by modelers. In this context the characterization of attractors, fixed points, and limit cycles is of interest.
3. What are the principles underlying cell-to-cell variability in the apoptosis response of a cell population? Why do cell types differ in their sensitivity to death-inducing stimuli? Currently, several therapeutic applications are tested to stimulate apoptosis in cancer cells, to decelerate tumor growth, or to prevent cells, preferentially neurons or cardiomyocytes, from undergoing programmed cell death. Modeling approaches could help to plan therapies and to predict the outcome on a population of cells. Particularly by distinguishing cell death

kinetics and the behavioral heterogeneity of different cell types, and predicting drug sensitization by cotreatments, modeling could be a valuable tool. We will describe and review strategies to predict cell death kinetics of single cells and of heterogeneous cell populations.

First we will give an overview of the basics of mathematical formalisms and then review successful application of ODE apoptosis models to resolve biological questions.

1.2 Overview of Mathematical Formalisms

Analyzing the cell on a systems view can be done by top-down and bottom-up approaches. Detailed mechanistic mathematical models constructed from the molecular characteristics of individual proteins ("bottom-up models") have only been developed for metabolic and signaling networks. In contrast, transcriptional regulatory networks, and the link between signaling networks and ultimate cellular decisions are best tackled by statistical methods which integrate huge amounts of data but are mostly phenomenological ("top-down modeling").

Top-down approaches examine the cell on a global level, treating individual regulatory modules as black boxes that are not analyzed mechanistically but only characterized with respect to input–output behavior. Thus, top-down methods typically do not require much prior knowledge about the system, so that many signaling and/or metabolic pathways can be studied at once. Most top-down approaches are solely data-driven and rely on high throughput screens of cellular behavior (gene expression profiling, proteomics, siRNA screening, sequencing, and affinity assays). Typically, the ultimate goal of top-down approaches is to identify biologically relevant patterns and correlations to the data (e.g., disease marker gene identification) or to predict new molecular interactions (e.g., reverse engineering algorithms).

Bottom-up approaches focus on well-characterized parts of the biochemical regulatory network, and are typically based on the assumption that the properties of these subnetworks (or "modules") can be studied in isolation. Based on prior knowledge and on time-resolved experimental data, mechanistic mathematical models describing the interactions of individual proteins in the module are constructed (e.g., by using sets of coupled differential equations). The goal of bottom-up modeling is to identify physiologically relevant systems-level properties emerging from complex interactions within the network (e.g., feedback).

Apoptosis-inducing signaling cascades, especially those induced by death ligands, were mainly studied using bottom-up modeling approaches, since (1) the molecular events are well characterized; (2) transcriptional events can be neglected; (3) the ultimate death decision often closely correlates with all-or-none activation of effector caspases, implying that statistical methods are not required to link signaling to cellular phenotypes. However, bottom-up approaches to apoptosis are diverse and the methodology of choice depends on the complexity of the signaling network under study, the available experimental data, and the question to be addressed by modeling. Boolean approaches are typically employed to qualitatively analyze the (quasi-)static behavior of large apoptosis-survival crosstalk networks which comprise many molecular species. Ordinary differential equation (ODE) models allow for the quantitative description of network dynamics but typically require knowledge about many kinetic parameters which either limits the network size and/or requires huge amounts of experimental data. Standard ODE modeling may even not be sufficient if spatiotemporally resolved single-cell data is available (1) spatial gradients within the cell can be modeled using subcellular compartment ODE models or partial differential equations (PDEs). (2) Cell-to-cell variability may arise due to stochastic dynamics of the apoptotic signaling cascade ("intrinsic noise") or due to cell-to-cell variability in the expression of pathway components ("extrinsic noise"). While ODE models with randomly sampled initial protein concentrations can be employed to simulate extrinsic noise, stochastic simulation algorithms are required to understand intrinsic noise. In the following, we will give an overview of top-down and bottom-up modeling approaches applied to apoptosis signaling, before discussing applications of ODE models in more detail.

1.2.1 Linear Regression Models

To systematically analyze how the pro- and anti-apoptotic cytokines TNF, EGF, and insulin impinge on the cellular apoptosis decision, Janes et al. (2005) generated a compendium of costimulation measurements. Based on the assumption that simple linear combinations of signaling activity profiles account for apoptosis initiation, they employed a top-down modeling approach known as partial least-squares regression (PLSR) which does not require prior knowledge. PLSR modeling calculates super axes as an orthogonal set of "principal components," which contain linear combinations of the original signaling protein activities weighted by their contribution to the apoptotic outputs. Thereby, the dimension of the data matrix is reduced to a small set of informative super axes, which can be used to predict apoptosis initiation for any experimental condition, provided that measurements of signaling species used for model training are available. PLSR has been successfully applied to other large-scale apoptosis datasets, and provided insights into complex phenomena such as autocrine amplification loops (Janes et al. 2006). For a more detailed description, please, see the chapter by Deppmann and Janes. A major drawback of PLSR is the lack of mechanistic insights into (1) how signaling activity patterns are generated and (2) how signaling activities are integrated, e.g., at the level of caspases, to control the death decision. Therefore, the next section will be devoted to bottom-up approaches applied to apoptosis which take into account mechanisms of apoptosis initiation.

1.2.2 Boolean Models

Recent biomedical research revealed a plethora of protein–protein and enzymatic interactions, and thus extensively characterized the topology of the intracellular signaling network. However, quantitative information characterizing the affinity of protein–protein interactions or enzyme kinetic parameters is still scarce. Moreover, quantitative characterization is often performed using recombinant proteins in vitro, with questionable relevance to the in vivo situation. Simulations of large-scale networks is therefore often performed using Boolean or logic modeling, a qualitative approach that is based on network topology, but does not take into account quantitative features of individual reactions. Instead protein activities are represented by nodes which can either be on or off (activity 0 or 1), depending on the activities of upstream input nodes. Logic rules are applied at each iteration: For example, in a so-called *AND-gate*, the node Z will be activated if and only if *both* input nodes *X* and *Y* are active. In contrast, an *OR-gate* simply requires either *X* or *Y* to be active. Thus, Boolean rules can be used to qualitatively represent real biochemical mechanisms such as functional redundancy (*OR-gate*) or coincidence detection (*AND-gate*), the latter, arising from sequential processing by two distinct enzymes. Since logical rules are applied iteratively, the approach can be used to study temporal phenomena such as adaptation. Moreover, Boolean networks can exhibit nonlinear dynamic phenomena such as oscillations, and stable vs. unstable attractors. Please see chapters by Schlatter et al. and Calzone et al. for a detailed review on Boolean network dynamics.

A number of Boolean modeling studies have been presented in the context of apoptosis (Calzone et al. 2010; Mai and Liu 2009; Philippi et al. 2009; Schlatter et al. 2009; Zhang et al. 2008). All these studies analyzed the crosstalk of apoptosis signaling via caspases and survival pathways such as NF-κB signaling. The main goal was the identification of stable states in the systems, representing cell fates such as apoptosis, necrosis, and survival. Calzone et al. (2010) and Mai and Liu (2009) focused on signaling upon death receptor engagement. They showed that the stable states of the apoptosis network are robust and investigated the requirements for irreversibility in the apoptosis decision. Schlatter et al. (2009) and Philippi et al. (2009) took into account costimulation with prodeath and prosurvival ligands, and experimentally confirmed key model predictions. Zhang et al. (2008) analyzed antigen-induced survival signaling network in T cell large granular lymphocyte (T-LGL) leukemia cells including transcriptional induction of cytokines and autocrine stimulation events. Model predictions could be confirmed in leukemic cells isolated from patients, thus contributing to our understanding of signaling deregulation in the disease. Taken together, Boolean modeling approaches provided valuable insights into apoptosis at multiple timescales and for various experimental settings.

1.2.3 Quantitative Modeling Approaches

Boolean models are inherently limited in their capability of quantitatively describing the temporal dynamics of biochemical networks. In the context of perturbation analysis, Boolean approaches are restricted to the simulation of complete elimination of network nodes and/or reactions; thus, gradual phenomena such as dosage compensation cannot be studied. Moreover, the qualitative effects of perturbations as revealed by Boolean modeling are often intuitively clear. Thus, in many cases, nontrivial and experimentally testable predictions require quantitative modeling approaches such as ODE and PDE modeling, as well as stochastic simulations.

ODE approaches, described in detail below, assume that large numbers of signaling molecules are present within the cell, so that random fluctuations in reaction events can be neglected by averaging over the whole molecule population. Moreover, in ODE modeling it is assumed that the cell represents a well-stirred reactor, implying that diffusion effects do not matter. In apoptosis networks, these assumptions are likely to be fulfilled, as caspase and their regulators are typically expressed at the number of several hundred thousand molecules per cell (Svingen et al. 2004). Furthermore, the time scale of apoptosis induction (hours) is slow relative to the time scale of protein diffusion within a cell (milliseconds to seconds); therefore, spatial gradients of apoptosis signaling molecules are unlikely to play a decisive role in apoptosis initiation.

Nonetheless, reaction–diffusion models allowed investigating molecular mechanisms of apoptosis induction: Using live-cell imaging with high temporal resolution, Rehm and colleagues (2009) observed that cytochrome c release from mitochondria during apoptosis occurs in spatial waves that propagate from a subcellular mitochondrial pool to the remainder of the mitochondrial population. PDE modeling was employed to investigate the dynamics of nonsteady state diffusion. This approach revealed that localized release and diffusion of inducers of mitochondrial outer membrane permeabilization (MOMP) alone was insufficient to explain the data. However, then the authors took into account that MOMP inducers bind to mitochondria, and modeling indicated that this absorption shapes the dynamics of cytochrome c release, thus providing insights into molecular mechanisms controlling apoptosis induction.

Owing to low molecule numbers of Bcl-2 family members, stochastic simulations using cellular automaton approaches were performed by Chen et al. (2007a), Siehs et al. (2002), and Düssmann et al. (2010) to describe the dynamics of MOMP. Chen et al. (2007a) focused on bistability and concluded that the stochastic system attained two distinct stable states much like the deterministic case; thus robustness of switching towards molecular noise could be confirmed. Düssmann et al. (2010) compared their model to measurements in cells expressing Bax-FRET probes monitoring Bax oligomerization. Their model could provide an explanation for pore formation upon Bax accumulation and oligomerization in the outer mitochondria membrane.

Live cell imaging tools are increasingly important and allow the analysis of apoptosis at the single cell level or even with subcellular resolution. Thus, stochastic and reaction–diffusion modeling are likely to become central to apoptosis modeling. For example, death receptors are frequently expressed at low levels and form localized (nano-) clusters on the cell membrane (Dumitru and Gulbins 2006), implying that deterministic ODE approaches will fail, especially upon weak stimulation. Stochastic and reaction–diffusion modeling will reveal underlying mechanisms and, more importantly, predict strategies for intervention for testing the functional relevance of such phenomena.

1.3 Basic Concepts in ODE Modeling

In this section, we give an overview of the most important steps in ODE model that include implementation, optimization, and analysis.

1.3.1 Building Blocks of Biochemical Models

The kinetics of chemical reactions can be described with reaction rates dependent on the concentrations of educts and products. Specifically, one typically assumes that the number of product molecules synthesized in a certain time interval is linearly dependent on the concentrations of educt molecules (law of mass action). The net influx or efflux arising from all participating reactions determines the rate of change in each molecular species. Thus, ODE modeling is based on the assumption that the temporal derivatives of molecule concentrations equal the sum of all relevant reaction rates (Table 1.1). Larger biochemical signal transduction networks are therefore reflected using coupled ODEs.

Table 1.1 lists elementary reactions in biochemical networks: Most steps in ODE models are described as unimolecular reactions that could represent irreversible reactions representing processes as degradation or substrate cleavage (1.1) or reversible transitions between certain states of a protein (1.2). Other common elements are the reversible assembly of two proteins, such as ligand binding to a receptor (1.3), reversible dimerization of two monomers (1.4), and enzyme–catalyzed reactions (1.5). In many cases, the full enzyme catalysis mechanism (enzyme + substrate $\leftrightarrow$ enzyme–substrate complex $\rightarrow$ enzyme + product) can be described by a single overall reaction rate, e.g., by using the Michaelis–Menten approximation (see biophysical textbooks).

Table 1.1 Exemplary components of a model graph

Unimolecular irreversible reaction	$A \xrightarrow{k} B$	$\frac{dA}{dt} = -kA$	(1.1)
Unimolecular reversible reaction	$A_1 \underset{k_-}{\overset{k_+}{\leftrightarrow}} A_2$	$\frac{dA_2}{dt} = k_+A_1 - k_-A_2$	(1.2)
Reversible ligand–receptor binding	$R \underset{k_-}{\overset{k_+}{\leftrightarrow}} RL$ (with L)	$\frac{dRL}{dt} = k_+R \cdot L - k_-RL$ $\frac{dRL}{dt} = 0 \Leftrightarrow K_b = \frac{k_+}{k_-} = \frac{RL}{R \cdot L}$	(1.3)
Dimerization	$2\,M \underset{k_-}{\overset{k_+}{\leftrightarrow}} D$	$\frac{dD}{dt} = k_+M^2$	(1.4)
Enzyme–catalyzed reaction	$A \xrightarrow{k} B$ (catalyzed by E)	$\frac{dB}{dt} = kAE$ $E = \text{const.}$	(1.5)

1.3.2 Simulation

Based on such simple building blocks, mechanistic models of biochemical reaction networks can be implemented. As a demonstrative example, we constructed a model of caspase activation by death ligands (Fig. 1.1a), where each reaction is described by an equation similar to those in Table 1.1. Signaling is initiated by reversible ligand binding to the death receptor, followed by formation of the so-called death-inducing signaling complex (DISC), recruitment of procaspase-8 and procaspase-3 cleavage by active caspase-8. The model consists of the stimulus (*L*; assumed to have a constant concentration in the medium), eight dynamical variables (*R*, *LR*, *DISC*, *C*8, *DISC*.*C*8, $C8^*$, *C*3, $C3^*$), and seven kinetic parameters (k_{1+}, k_{1-}, k_2, k_{3+}, k_{3-}, k_4, k_5). The kinetic parameters and initial concentrations were taken from previous theoretical and experimental studies (Albeck et al. 2008b; Bentele et al. 2004; Neumann et al. 2010; Rehm et al. 2009; Stennicke et al. 1998).

Using numerical integration techniques, the temporal evolution of the model species to extracellular stimulation by death ligands can be simulated (Fig. 1.1b). The simplest numerical integration method, known as the Euler method, approximates the solution of the differential equation $dx/dt = f(x)$ iteratively by the discretization

$$x(t_{i+1}) = x(t_i) + f(x(t_i)) \cdot \Delta t, \tag{1.6}$$

where $x(t_i)$ is the solution at time point t_i and $\Delta t = t_{i+1} - t_i$. In practice, the solution of the differential equation at any time point is obtained by iteratively applying (1.6) starting from the initial conditions at $t = 0$. The smaller the time increment Δt is chosen, the more accurate solution might be obtained. However, in general, the numerical error of the Euler method increases with increasing number of

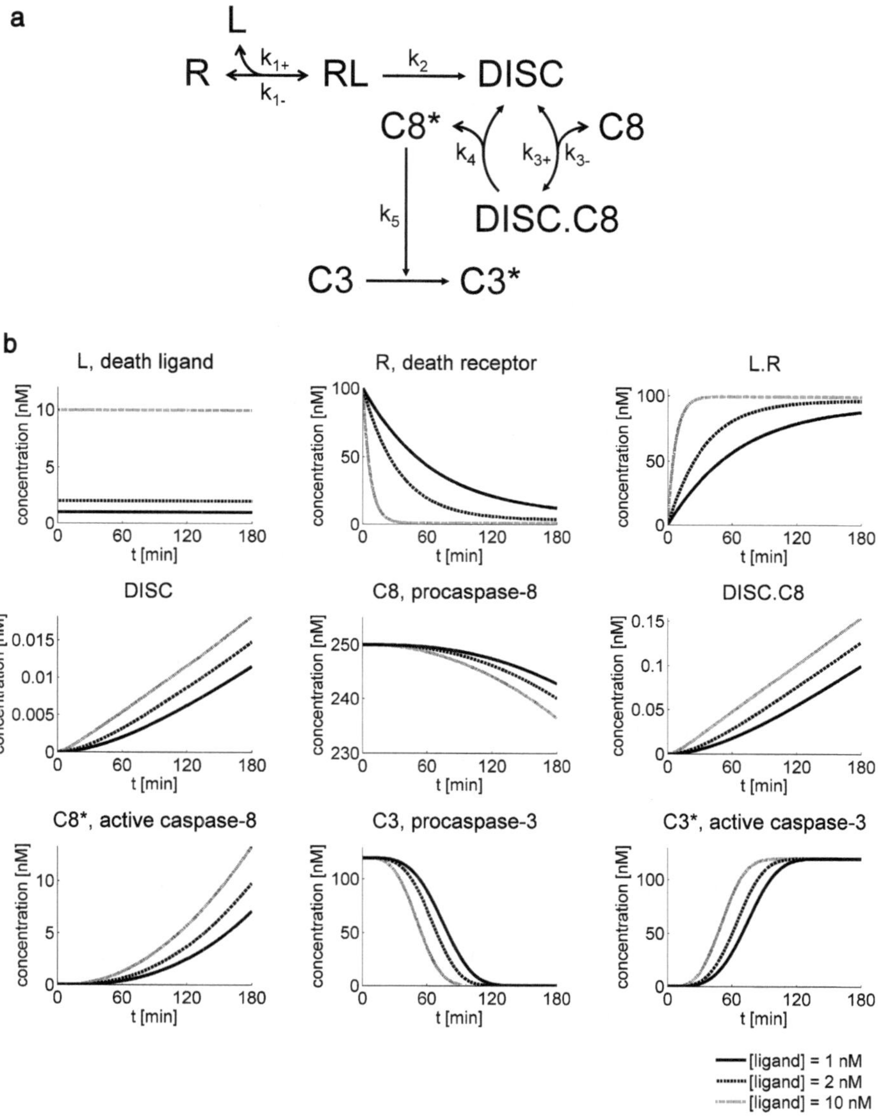

Fig. 1.1 Exemplary model of extrinsic apoptosis and predicted trajectories for its variables. (**a**) The model graph represents five reactions that are translated into a set of eight ODEs. The ligand in the medium is assumed to be present in excess, and is therefore not described by a differential equation, but considered to be constant. The model variables are receptor (R with $R_0 = 100$ nM), receptor ligand complexes (*RL*), DISCs, procaspase-8 (C8 with $C8_0 = 250$ nM), procaspase-8 bound to DISCs (*DISC.C8*), active caspase-8 (C8*), procaspase-3 (C3 with $C3_0 = 120$ nM), and active caspase-3 (C3*). (**b**) Simulated model trajectories for a step-like increase in the death ligand stimulus (see legend for ligand concentrations)

numerical integration steps. Thus, more accurate algorithms are usually applied, e.g., the Runge–Kutta method which uses a weighted average of slopes instead of a single slope $f(x(t_i))$ (see textbooks on ODEs). Corresponding numerical integration algorithms are integrated in standard mathematical computing software packages (e.g., Matlab, Maple, Mathematica).

The simulations in Fig. 1.1b reveal that the concentrations in the receptor–ligand module no longer change at the later time points, indicating that this subnetwork has reached (quasi-) steady state. At steady state the influxes and effluxes of each molecule even out, implying that the receptor–ligand system fulfills $\mathrm{d}RL/\mathrm{d}t = 0$. Steady states represent important characteristics of biochemical reaction networks and are amenable to mathematical analysis, since the differential equation system simplifies to a set of algebraic conditions. In the context of cellular signaling, steady states are often analyzed to show that the system can reversibly switch between two distinct states, or undergo an irreversible switch as in the case of cell death (see Sect. 1.4).

1.3.3 Parameter Estimation, Identifiability, and Model Selection

In many cases, the model parameters p_j comprising initial concentrations of model species and kinetic constants are not known, but needed to be estimated by fitting the model to experimental data using optimization algorithms. One problem in fitting is that the model is formulated in absolute concentrations of the species, while experimental data in biology can often only be obtained in relative units (e.g., relative intensities of immunoblot bands). In this case, the model trajectories are multiplied by a fitted scaling factor to match the experimental measurements. Thus, by introducing another degree of freedom, only the shape of the model trajectory is adjusted to the data, but not its absolute value(s). The parameters are chosen in such a way that they optimize the correspondence between the measurements E_{ij} of the m experimentally species at times t_j for $j = 1 \ldots n$ with standard deviations σ_i and the corresponding simulated values S_i. Most often the maximum likelihood estimator

$$\chi^2 = \sum_{i=1}^{m} \sum_{j=1}^{n} \left(\frac{E_{ij} - S_{ij}}{\sigma_{ij}} \right)^2, \tag{1.7}$$

is used for this purpose (Press et al. 1992; Raue et al. 2009; Timmer et al. 2004). A reasonable model fit to N data points in total is obtained if the condition $\chi^2 < N$ holds, indicating that the model is on average within the experimental error. Several algorithms for parameter estimation have been proposed including simulated annealing, genetic algorithms, and the Levenberg–Marquardt method (Levenberg 1944; Marquardt 1963). The maximum likelihood estimator (1.7) of the fitted models can be used to discriminate between different model topologies: Specifically, the presence or absence of certain biochemical mechanisms can be verified or neglected by comparing model variants with respect to their ability to fit the experimental data. In particular, one often compares so-called nested model variants, where the larger variant includes all reactions of a smaller one. Model selection approaches such as the Akaike information criterion and the likelihood ratio test weigh the goodness of fit (χ^2) with the number of model parameters, and allow the discrimination of nested models. Specifically, the larger model (which includes the topology of the smaller model) is rejected in case it only improves the fit within experimental error ("overfitting"). A major problem in parameter estimation is nonidentifiability of parameters: typically, not all parameters can be unambiguously determined from experimental data ("nonidentifiability"). This leads to uncertainties in model predictions. Tools have been developed to calculate (1) which parameters can be identified from the data and (2) confidence intervals for model predictions (Raue et al. 2009, 2010). In the context of model discrimination, experimental conditions can be predicted where the distance between simulated trajectories that result from different topologies is maximal; these conditions are then used to experimentally discriminate between model topologies, and to further refine the parameter estimation results (iterative cycle between experiment and model-based experimental design).

1.3.4 Sensitivity Analysis

Once a realistic model has been established, it might be important to identify points of fragility in the network to guide model reduction, experimental design or pharmacological intervention, and to understand robustness principles. The behavior of a signaling system described with a set of coupled ODEs changes when initial concentrations of the model species or kinetic constants are altered. Sensitivity analysis systematically quantifies such responses by calculating how the concentration of the model species c_i reacts to alterations in each parameter p_j.

$$s_{ij} = \frac{\partial c_i/c_i}{\partial p_j/p_j}. \tag{1.8}$$

The sensitivities s_{ij} (1.8) are often calculated at a steady state. To characterize time-dependent sensitivity, Bentele et al. (2004) introduced an absolute sensitivity S_{ij} (1.9) which sums up the sensitivities over the whole trajectories.

$$S_{ij} = \int |s_{ij}| \mathrm{d}t. \tag{1.9}$$

In the case of programmed cell death, the ligand concentration needs to exceed a certain threshold to induce cell death. Sensitivity analysis provides a useful tool to investigate how this threshold and thus cellular susceptibility to apoptosis could be modulated pharmacologically (Hua et al. 2005; Legewie et al. 2006). Moreover, sensitivity analysis has been applied to identify reactions determining whether a cell dies by type I or type II cell death (Okazaki et al. 2008), and to reduce model complexity (Bentele et al. 2004).

In the theoretical study of Aldridge et al. (2006) another method based on direct finite–time Lyapunov exponents (DLEs) was presented to provide an extension to common sensitivity analysis methods. This method is used to analyze dynamic changes in an apoptosis signaling network upon simultaneous perturbations of several model components. The paradigm that the decision of a cell towards or against apoptosis does not depend on the concentration of a single signaling protein but rather is defined by a combination of several initial concentrations also motivated another theoretical study (Hua et al. 2006). An ODE model was used to simulate model trajectories for various initial protein concentrations, and a decision tree was constructed reflecting outcomes for given initial concentration sets by using the *classification and regression trees* method (Breiman et al. 1984). Thus, the outcome to an apoptotic stimulus could be predicted depending on a set of relative signaling protein concentrations with a certain accuracy.

1.4 Conceptual Model of an Irreversible Switch in Apoptosis

Programmed cell death should be initiated in all-or-none manner, as incomplete cellular demise may support the development of diseases such as cancer. Accordingly, single-cell measurements revealed that caspase activation occurs in a digital manner (Albeck et al. 2008b). In this section, we will derive a minimal model of apoptosis initiation by caspases which exhibits two stable steady states ("bistability") that show very different caspase activation levels, corresponding to life and death.

Consider the simplified topology of extrinsic apoptosis (Fig. 1.2a), which does not take into account details of CD95L binding to its receptor and DISC formation, but describes all these steps by a single reaction step. Accordingly, the differential equation for caspase-3 (1.10) comprises a linear term f, which describes ligand-induced cleavage of procaspase-3 into caspase-3 (1.11). Active caspase-3 is known to enhance its own production as a part of a positive amplification loop, for example, by enhancing caspase-8 activation or cytochrome c release from mitochondria. We assume that caspase-3 feedback exhibits *cooperativity*, implying that increasing the concentration of $C3^*$ facilitates further caspase-3 activation in a highly nonlinear manner. In the model, we described feedback by a *Hill equation* term g (1.12), which shows a steep, cooperative response for increasing values of the exponent m. The term h (1.13) describes the degradation of caspase-3.

$$\frac{dC3^*}{dt} = f + g(C3^*) + h(C3^*), \tag{1.10}$$

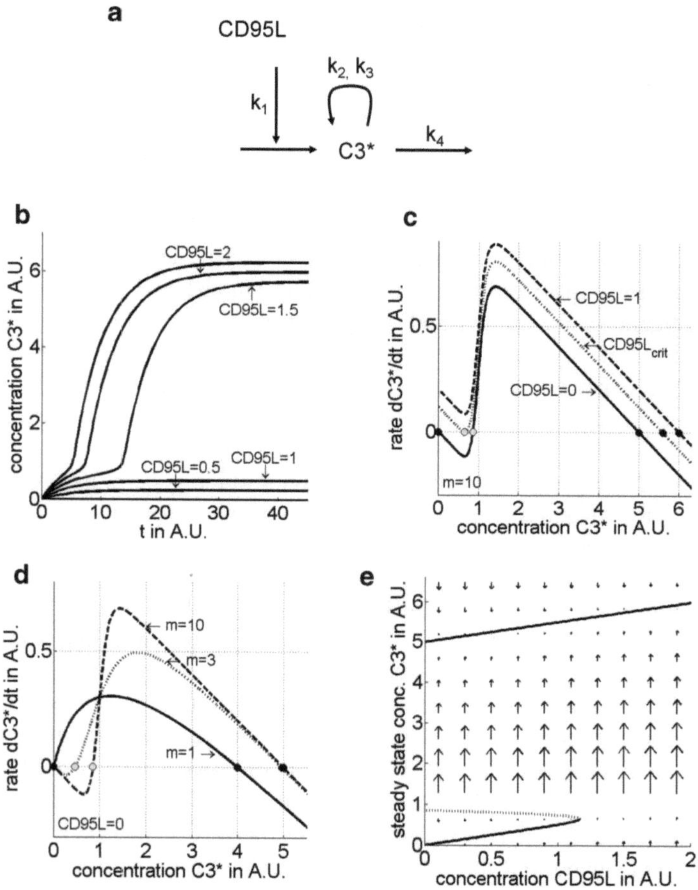

Fig. 1.2 Switch-like activation of caspase-3 arising from positive feedback and cooperativity. (**a**) Model graph of CD95 ligand (CD95L) dependent activation, cooperative self-activation, and degradation of active caspase-3 (C3*) with kinetic constants k_1–k_4. (**b**) Concentration time series of active caspase-3 at different concentrations of the CD95 ligand with parameters $k_1 = 0.1$, $k_2 = k_3 = 1$, $k_4 = 0.2$, and $m = 10$. Above a critical concentration of CD95L, a switch-like increase of the concentration of active caspase-3 occurs. (**c** and **d**) Temporal derivative of active caspase-3 dependent on the concentration of caspase-3 for different concentrations of the CD95 ligand at the same value for exponent m (**c**) and for different exponents m (**d**) representing no ($m = 1$), intermediate ($m = 3$) or strong cooperativity ($m = 10$) at the same CD95L concentration. Stable steady state concentrations of C3* are denoted with filled black circles, unstable steady states with empty circles. (**e**) Signal response diagram of caspase-3 activation. The *solid* and *dashed lines* represent stable and unstable steady states of caspase-3, respectively. Upon increase of CD95L concentrations when a certain ligand concentration is reached the lower of the previously two first steady states vanishes and the system abruptly switches to the remaining higher steady state concentration. Even after removal of the stimulus this higher steady state is not left representing an irreversible switch. *Arrows* denote locations of hypothetic trajectories and their directions from certain start values of C3* and CD95L. Their lengths correspond to $dC3^*/dt$

$$f = k_1 \mathrm{CD95L}, \tag{1.11}$$

$$g(C3^*) = k_2 \frac{C3^{*m}}{C3^{*m} + k_3}, \tag{1.12}$$

$$h(C3^*) = -k_4 C3^*. \tag{1.13}$$

In Fig. 1.2b time series of active caspase-3 at different concentrations of the CD95L are shown that result from this exemplary model structure. Above a certain threshold ligand concentration and after a lag phase, $C3^*$ increases abruptly to a higher steady state level as observed experimentally at the single cell level (Albeck et al. 2008b; Rehm et al. 2002). In this model, trajectories that reach higher $C3^*$ levels could be regarded as "cell death" and such that remain at a lower level as "survival."

Figures 1.2b, c illustrate the dynamic behavior of the system. Figure 1.2b shows the trajectories of active caspase-3: low ligand concentrations do not induce a remarkable $C3^*$ activation; above a critical CD95L concentration the system switches to higher $C3^*$ concentrations. At all ligand concentrations, the system reaches a steady state after sufficiently long times (Fig. 1.2b). Figure 1.2c shows the dependency of the rate $dC3^*/dt$ on the caspase-3 concentration, and provides insights into the system steady states which fulfill the condition $dC3^*/dt = 0$. In particular, the irreversible nature of apoptosis initiation can be understood: In the absence of stimulation (CD95L = 0), the model system exhibits a steady state with no caspase activation ($C3^* = 0$); this life steady state is stable, since for small perturbations ($C3^* > 0$), the caspase-3 pool will decay ($dC3^*/dt < 0$). Moreover, the system exhibits two non-zero steady states ($C3^* = 1$; $C3^* = 5$) even in the absence of ligand, but these can only be reached if caspase-3 was previously activated strongly (remember that the caspase-3 pool will monotonously decay for $C3^* < 1$). The intermediate steady state is unstable ($dC3^*/dt < 0$ for $C3^* < 1$ and $dC3^*/dt > 0$ for $C3^* > 1$), while the upper steady state is stable ($dC3^*/dt > 0$ for $C3^* < 5$ and $dC3^*/dt < 0$ for $C3^* > 5$). The existence of a stable death state even in the absence of CD95L implies that our model captures the irreversible nature of apoptosis induction.

What are the requirements to reach this death state by external ligand stimulation? Does the system exhibit a sharp threshold? Increasing the CD95L concentration shifts the life steady state to low, nonzero caspase-3 activation levels. Once CD95L exceeds a critical concentration ($\mathrm{CD95L}_{crit}$), the lower two steady states suddenly disappear, and the system shifts to the death state ($dC3^*/dt > 0$ for $C3^* < 6$ in case of CD95L = 1). Thus, a sudden, qualitative change in the system behavior is observed at $\mathrm{CD95L}_{crit}$, and this phenomenon is known as a bifurcation. In biological terms, all-or-none caspase-3 activation from nonapoptotic to apoptotic levels occurs at the critical CD95L concentration. Once the system passed the threshold, caspase-3 will be activated irreversibly even if the ligand is removed.

Figure 1.2d illustrates the relevance of the exponent m for the existence of steady states. In the absence of cooperativity ($m = 1$), $C3^*$ exhibits only one stable steady state concentration, thus implying gradual and reversible caspase activation. This demonstrates that *nonlinear, sigmoidal* positive feedback is essential for all-or-none and irreversible apoptosis initiation.

Figure 1.2e shows the steady states of the system as a more common dose–response diagram of caspase-3 activation. The stable steady states are indicated as solid black lines, while the unstable steady state is shown as a dashed line. Moreover, the flow of the system ($dC3^*/dt$) is indicated for any initial $C3^*$ concentration (arrows). As expected, this representation includes the same features as Fig. 1.2c: Starting from a low ligand concentration the system is kept at low $C3^*$ concentrations. At a critical CD95L threshold concentration, the lower steady states disappears, and $C3^*$ switches in an all-or-none manner to a high concentration level. The unstable steady state (dashed line) represents a so-called separatrix of the system: the concentration of $C3^*$ decreases below the unstable steady state, while it increases otherwise. This implies that caspase-3 activation in our model is an irreversible process; once the higher steady state has been reached removing the ligand is insufficient to cross the separatrix, and the high steady state remains. We conclude that our model can reproduce the experimental observation that apoptosis initiation is an irreversible process characterized by a point-of-no-return.

Several more realistic apoptosis models, reviewed in the next section, also incorporate highly nonlinear positive feedback and bistability, and are thus based on the same building blocks as our conceptual model. The models mainly differ in the molecular mechanisms responsible for bistability of apoptosis, e.g., caspase activation (e.g., Eissing et al. 2004; Legewie et al. 2006; Bagci et al. 2006), interactions of the Bcl-2 protein family members (Cui et al. 2008), trimerization of death receptors (Ho and Harrington 2010).

1.5 Mechanistic ODE Models Describing Apoptosis Networks

The conceptual model of apoptosis presented in Sect. 1.4 provided insights into mechanisms of all-or-none and irreversible decision making, but is certainly an oversimplification. In this section, we will review other more complex models of apoptosis.

The first mechanistic apoptosis model of coupled differential equations, presented by Fussenegger et al. (2000), described sequential activation of caspase-8, caspase-9, and caspase-3 by intrinsic or extrinsic stimuli. The model encountered for a positive feedback from caspase-3 promoting the release of cytochrome c from mitochondria and thus promoting the additional activation of caspase-9 in apoptosomes. As this model was not trained against quantitative

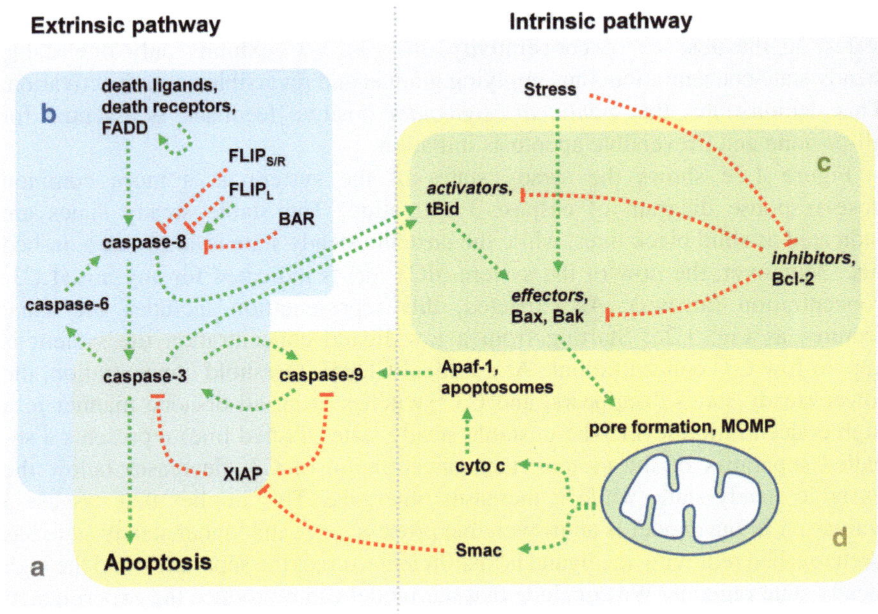

Fig. 1.3 Overview of the extrinsic and intrinsic parts of the apoptotic signaling network. Subareas are indicated that were investigated by current modeling studies. (**a**) Modeling studies that captured extrinsic and intrinsic pathways (Albeck et al. 2008b; Bagci et al. 2006; Bentele et al. 2004; Fussenegger et al. 2000; Harrington et al. 2008; Hua et al. 2005; Okazaki et al. 2008). (**b**) Modeling studies with main focus on the level of death receptors, DISC assembly, and caspase activation processes (Eissing et al. 2004; Fricker et al. 2010; Ho and Harrington 2010; Neumann et al. 2010; Würstle et al. 2010). (**c**) Modeling studies that described interactions between members of the Bcl-2 family prior to mitochondria outer membrane permeabilization (MOMP) (Chen et al. 2007a, b; Cui et al. 2008; Düssmann et al. 2010; Siehs et al. 2002). (**d**) Modeling studies with main focus on intrinsic apoptosis and caspase inhibition (Legewie et al. 2006; Rehm et al. 2006; Stucki and Simon 2005; Zhang et al. 2009)

data, it provided only predictions on activated fractions of initiator and executioner caspases dependent on initial concentrations of apoptosis promoting or inhibiting proteins. A lot of progress has been made since this first apoptosis model, and different aspects have been studied in detail.

In the following, we will review the current literature on ODE-based apoptosis modeling. Figure 1.3 shows an overview about subareas of the apoptotic signaling network that were investigated by current models. First, we will describe models which are similar to the conceptual model in the sense that they understand apoptosis as a bistable process. Second, we will summarize studies investigating the temporal dynamics of apoptosis. Third, crosstalk models describing apoptosis and survival networks will be discussed. The final focus of our review will be cell-to-cell variability.

1.5.1 Origins of a Robust All-or-None Behavior: Models Characterized by Bistability and Feedback Mechanisms

Bistability may play an important role for all-or-none and irreversible decision making, thus allowing the suppression of noise and prevention from accidental apoptotic stimuli. In the following, we will review models exhibiting bistability due to positive feedback in the intrinsic or extrinsic apoptosis pathways.

1.5.1.1 Extrinsic Apoptosis Pathway

Understanding bistability in the process of apoptosis initiation was the focus of the study of Eissing et al. (2004). Their model described the bistability in extrinsic apoptosis within the context of caspase-mediated positive feedback. Caspase-8 activated by receptor-induced apoptosis in type I cells activates caspase-3, while caspase-3 promotes positive feedback by caspase-8 activation. A stability analysis of this minimal model showed that bistability and therefore a stable live steady state were only possible parameter values far off the experimentally measured kinetic parameters. By extending the model topology, the authors concluded that bistable caspase activation within the physiologically reasonable parameter range required the consideration of inhibitors of activated caspase-8. Specifically, it was suggested that caspase-8 inhibitory proteins such as bifunctional apoptosis regulator protein (BAR) (Zhang et al. 2000), and caspase-8 and -10-associated RING proteins (CARPs) (McDonald and El-Deiry 2004) play a central role for establishing bistability. The important antiapoptotic role of the protein BAR was investigated in the study of Pace et al. (2010).

Bistability on the ligand/receptor level was proposed upon theoretical considerations of a positive feedback in receptor oligomerization reactions (Ho and Harrington 2010). These were motivated by new insights into the structure and function of CD95 (APO-1/Fas) molecules (Scott et al. 2009). Protein crystallization experiments on receptor/FADD-clusters had shown that receptors in the absence of ligands favor a closed form where FADD cannot bind. Upon ligand binding an open form is favored, allowing FADD binding, DISC formation, and signal progression. The conformation of open receptors allows that open receptors stabilize each other, which causes receptor oligomerization and positive feedback amplification. An ODE model was formulated in a continuum approximation where molecule numbers were treated as local protein concentrations. At high receptor densities, which could be potentially induced locally inside lipid rafts, reactions can take place, where several open receptors stabilize each other. In the model these events could be approximated by higher-order reactions. As described in Sect. 1.4, rate equations for reactions that contain terms with an order of three or larger can have two stable steady states and one unstable steady state in between, which causes bistable behavior. Therefore, depending on the local receptor densities reversible or irreversible bistability can result. This leads to an all-or-none response on the level of DISCs, resulting in a gradual response, integrated over all clusters on the cell level.

1.5.1.2 Intrinsic Apoptosis Pathway

The theoretical study by Bagci et al. (2006) addressed origins of bistability on the level of MOMP and apoptosome formation. Two positive feedback mechanisms contribute to bistability: First, caspase-3 cleaves and inactivates the MOMP inhibitor Bcl-2, and thereby amplifies its own production. A second feedback arises from the cleavage of the MOMP inducer Bid by caspase-3; thus Bid cleavage, initially triggered by caspase-8, can be enhanced by caspase-3. Their mass-action model describes oligomerization of Apaf-1 bound to cytochrome c to the heptameric complexes of apoptosomes. As this cooperative oligomerization process leads to higher-order terms in the corresponding rate equation, the positive feedback interactions can result in bistable regimes corresponding to either survival or apoptosis. As the model of Bagci et al. (2006) does not consider reactions upstream of initiator caspases, initial doses of caspase-8 and caspase-3 serve as stimulus. Their considerations were motivated by experimental studies that had shown a survival mechanism of cancer cells based on elevated Bax degradation (Li and Dou 2000), decreased Bax expression in human breast cancers (Schorr et al. 1999), or overexpression of Bcl-2 (Reed 1999). In the model of Bagci et al. (2006) bifurcation points in the caspase-3 activity were investigated, that are dependent on the production or degradation of Bax and their relation to inhibitory Bcl-2 proteins. Above a certain threshold for the degradation rate of Bax or below a certain threshold for the Bax production rate, the bistable behavior is changed into a monostable survival state. In this state an initiator caspase stimulus cannot trigger the apoptosome-mediated feedback anymore. Recent studies on Bcl-2 family members show even more complicated relations on the level of mitochondria among proteins that act as sensitizers (as Bad, Noxa, or Puma), activators (Bid, Bim), or effectors (Bax, Bak). To characterize the vulnerability of tumors to apoptosis effectors, the impacts of different Bcl-2 family members on mitochondria isolated from tumor samples were investigated (Certo et al. 2006; Deng et al. 2007). For more detailed reviews see (Brunelle and Letai 2009; Chonghaile and Letai 2008; Vo and Letai 2010).

In a subsequent study the model was expanded to investigate the modulation of bistable switching by nitric oxide (NO) signaling which plays a dual role in the regulation of apoptosis and survival (Bagci et al. 2008): Several nitric oxide species such as NO or dinitrogen trioxide (N_2O_3) prevent apoptosis by inactivating both caspase-8 and caspase-3 (Li et al. 1997; Mannick et al. 1999; Rössig et al. 1999). In contrast, the NO species peroxynitrite (ONOO-) promotes apoptosis by triggering mitochondrial pore formation (Vieira et al. 2001). The model of Bagci et al. (2008) quantitatively describes the metabolism of nitric oxide species, and their effect on apoptosis. While NO species that exclusively inhibit caspase-8 only delay apoptosis, such species that inhibit caspase-3 as well as caspase-8 can prevent the bistable feedback from apoptosome formation, and cause survival. Bagci et al. also investigated the impact of glutathione which acts as a regulator of nitric oxide metabolism (Hu et al. 2006). Their model suggested that the proapoptotic effect of glutathione by inhibiting antiapoptotic NO species is stronger than its antiapoptotic effect by inhibiting NO species that facilitate mitochondria permeabilization.

1.5.1.3 Implicit Feedback Mechanisms in the Intrinsic Apoptosis Pathway

In the modeling studies summarized so far, the positive feedback mechanisms known from the biomedical literature and their contribution to bistability were analyzed. Additionally, mathematical modeling could provide valuable insights into non-obvious, hidden feedback loops that arise from the topology of the apoptosis network. This phenomenon has been referred to as implicit positive feedback regulation. The interplay of caspase-3, caspase-9, and inhibitors of apoptosis (IAPs) in the mitochondrial proapoptotic pathway was investigated in a model by Legewie et al. (2006): Cytochrome c released from mitochondria, triggers activation of caspase-9, which in turn cleaves procaspase-3 into active caspase-3. Both caspase-3 and caspase-9 are inhibited by XIAPs to prevent autoreactive activation. Interestingly, an implicit positive feedback loop arises from the dual inhibition of both caspases by XIAPs: Once active caspase-3 is generated, it can bind to XIAPs, thus sequestering XIAPs away from caspase-9. This sequestration effect enhances caspase-9 activation, resulting in autoamplification of caspase-3 cleavage. In the model by Legewie et al., the dependency of the concentration of active caspase-3 as the response to an Apaf-1 concentration shows different characteristics of either monostable, bistable reversible, or bistable irreversible behavior. The authors concluded that implicit positive feedback alone brings a very small range of bistability; however, implicit feedback synergizes with other feedback mechanisms to establish a broad bistable range and irreversibility in the life–death decision.

The studies of Chen et al. (2007a, b) combined ODE, stochastic and cellular automaton modeling to further understand signaling processes that potentially lead to MOMP. In these studies interactions between pore-forming *effectors* (Bax, Bak), *activators* and *enablers* (tBid and several others), and *inhibitors* (Bcl-2 amongst others) that lead to or prevent mitochondria outer membrane permeabilization are analyzed. After translocation to mitochondria, inactive Bax and Bak are catalyzed to their active form by an *activator*. Subsequently, activated Bax and Bak lead to membrane pore formation and cell death. In the study of Cui et al. (2008), models involved in the bistability of MOMP were further developed. Questions on the possible model topology led to hierarchical considerations in the studies of Chen et al. (2007a, b) as well as Cui et al. (2008): Do *activators* and *enablers* indirectly induce apoptosis by sequestering Bcl-2 away from Bax, or are activators directly proapoptotic by catalyzing the reaction of Bax to its active form that can cause pore formation and cytochrome c release? This question led to an indirect model recapturing the inhibition of Bcl-2 by *activators* and topologies describing direct Bax activation. *Inhibitors* as Bcl-2 in the indirect topology interfere by inhibiting Bax and thereby preventing its oligomerization at the pores. In direct topologies they inhibit *activators* from catalyzing Bax activation. Direct topologies were favored in this study, as they involve two possible feedback mechanisms that could contribute to a bistability in Bax activation. These considerations were motivated by experimental studies that showed a bimodal distribution of cells that had low or high amounts of activated Bax and Bak monitored by flow cytometry (Fischer et al. 2004; Gómez-Benito et al. 2005; Willis et al. 2005). The first described feedback mechanism is facilitated by activated Bax that can sequester Bcl-2, leading to an increase of free activators and thereby providing increased Bax activation. This mechanism is similar to the role of XIAP as discussed in the study of Legewie et al. (2006), since Bcl-2 acts as a dual inhibitor of upstream activators and their downstream effector Bax. Furthermore, the model of Cui et al. (2008) considers a feedback mechanism, in which activated Bax itself provides further Bax activation. In their study the signal response behavior is characterized by the dependency of active Bax and Bcl-2 steady state levels as dependent on the production rate of *activators*. In a model containing both feedback mechanisms the interval of activator production rates that lead to bistability of active Bax and Bcl-2 concentrations is significantly enlarged compared to a variant with only one feedback mechanism. Therefore, the combination of both feedback mechanisms would provide a higher robustness for the bistable behavior of Bax activation and mitochondria pore formation.

1.5.2 Origins of a Robust All-or-None Behavior: Switching and Threshold Mechanisms Other Than Bistability

Biochemical signaling networks may exhibit switching mechanisms other than bistability arising from positive feedback. In such cases, the system exhibits a single steady state which increases in a steep, nonlinear manner with increasing input concentration. Such sigmoidal, all-or-none dose–response behavior has been termed ultrasensitivity. One ultrasensitivity mechanism with particular relevance to apoptosis is inhibitor ultrasensitivity: here, a protein inhibitor strongly binds to its target, implying that the target remains completely inactive unless the total concentration exceeds the total inhibitor concentration (Ferrell 1996). Thus, the stoichiometry between inhibitor and target determines the system behavior, explaining why the mechanism is also known as stoichiometric switch.

The critical roles of c-FLIP$_L$ and c-FLIP$_S$, which potentially act as stoichiometric inhibitors in the DISC, were investigated in a model of Bentele et al. (2004). The dependence of the ligand concentration threshold on the concentrations of both splicing variants of c-FLIP was characterized, and it was concluded that c-FLIPs establish a stoichiometric switch. A large-scale model comprising DISC assembly, caspase activation, MOMP, interference from caspase inhibitors, and degradation processes was derived. The model could be fitted to quantitative Western blot data (caspase-8, -2, -7, -3, -9, Bid, PARP) representing population measures of protein concentrations under different ligand concentrations using hierarchical parameter estimation. By a global sensitivity analysis, clusters of modeled signaling proteins with high mutual sensitivities of protein concentrations were defined, which lead to functional subsystems. By disregarding parameters that had low sensitivities to parameters in one cluster, the dimensionality of the parameter estimation problem could be decreased. Predictions of the reduced model were subsequently verified experimentally. Most importantly, it could be shown that the threshold ligand concentration was highly sensitive to the c-FLIP concentration which is consistent with a stoichiometric switch mechanism. A refined version of the Bentele model was presented in the theoretical study of Toivonen et al. (2011) which took into account fast turnover of c-FLIP variants that could be relevant for their antiapoptotic effect. It could be shown that the concentration of c-FLIP at the time of ligand addition is central to apoptosis timing.

Another ultrasensitivity mechanism with potential relevance to apoptosis is protein dimerization. For protein dimerization the steady state of the active dimer depends on the total protein concentration in a quadratic manner as described in Sect. 3.1, (1.4). Thus, weak input signals controlling the concentration of the monomer species can be suppressed, while stronger inputs are transmitted. This phenomenon, known as multistep ultrasensitivity, was analyzed by Würstle et al. (2010) as described in the following.

In the absence of death receptor ligands the amplification loop from active caspase-8 to caspase-3, from activated caspase-8 to caspase-6 and back to caspase-8, has to be silenced to prevent apoptosis induction at low levels of active caspase-8. Würstle et al. (2010) compared different model variants to understand how efficient suppression of the amplification loop can be achieved. First, they analyzed the role of well-known caspase inhibitors such as the caspase-8 inhibitor BAR and the caspase-3 inhibitor XIAP. Second, they took into account that only caspase-8 dimers are catalytically active, and considered dedimerization of caspase-8 complexes released from the DISC. In this context, it had been shown experimentally that the caspase-8 dimerization equilibrium favors the formation of caspase-8 monomers (Pop et al. 2007). A core model describing the activation loop only was extended by either one of these three inhibiting mechanisms. Time series of caspase activation and substrate cleavage by caspase-3 caused by mild initial stimuli of caspase-8, -3, or -6, were calculated for each model variant, using experimentally measured kinetic constants for caspase activities and caspase-8 dimerization. Subsequently, time courses of the model variables were calculated under various initial concentrations of procaspases-3, -6, and -8, BAR, and XIAP to assess the sensitivity of the system towards each inhibitory mechanism in a time interval of 24 h, respectively. Each parameter constellation leading to less than 20% substrate cleavage was classified as non-apoptotic, while more than 80% substrate cleavage was considered as effective apoptosis. Thereby model variants could be compared regarding their potential to prevent apoptosis. A model was considered as more preventive if it caused weak cleavage in a larger fraction of the randomly chosen parameter sets than other model variants. As a result of an initial stimulus of caspase-8 the numbers of parameter constellations leading to hypothetic survival were slightly higher in the XIAP model than for the dedimerization model, while the number was highest for the dedimerization model in response to stimuli of caspase-3 or caspase-6. The apoptosis-preventing effect of the low affinity in a caspase-8 dimer becomes evident when considering that caspase-6 can only cleave monomers of procaspase-8 in absence of a dimerization inducer as the DISC. Because of the low affinity in the caspase-8 dimer, the system is especially stable against a stimulus of caspase-8 monomers. Consequently all three inhibitory mechanisms were included into one model to assess thresholds of the maximal caspase-8 stimulus strength that could be compensated. Threshold changes upon removal of one of the mechanisms were determined that showed again the strong perturbation resistance by caspase-8 dimerization and dissociation (Würstle et al. 2010). Thus, it was concluded that the caspase-8 dimerization equilibrium efficiently prevents accidental cell death initiation.

Another monostable model of the apoptosis threshold was introduced in the theoretical study of Stucki and Simon (2005). However, these authors did not focus on the mechanism of ultrasensitivity, but represented all-or-none caspase-3 activation phenomenologically using a Heaviside function in the caspase-3 production term. The major focus of the study was to analyze how the caspase-3 activation threshold could be modulated by the caspase-3 inhibitory XIAPs, the XIAP antagonist Smac, and Smac-binding antiapoptotic proteins such as survivin (Song et al. 2003). The potentially limiting role of caspase-3 degradation was addressed, and it was concluded that XIAPs efficiently suppress apoptosis by triggering the degradation of caspase-3 in a nonlinear manner.

1.5.3 Models Characterized by a Timing Switch

Steady states, bistable switches, and ultrasensitivity govern long-term decision making within biochemical signaling networks. However, in the context of apoptosis, it is also important that the time course of effector caspase activation is abrupt. Such temporal switching ensures complete and coherent initiation of cellular demise. Single cell measurements using GFP-tagged cytochrome C and caspase FRET probes confirmed that mitochondrial permeabilization and subsequent effector caspase activation indeed occur in a temporally abrupt manner (Goldstein et al. 2000; Rehm et al. 2002). Accordingly, a more recent study concluded that the apoptosis timing in single cells consists of a variable lag time followed by the sudden switch-like effector caspase activation (Albeck et al. 2008a). While the lag time varies within the range of one to several hours, dependent on the stimulus strength, the sudden switching time was shown to be relatively invariant around 30 min. This robustness of sudden switching can be interpreted as necessary to prevent from states of partial destruction that could cause genomic instability. The lag time is lengthened by proteins upstream of activated Bax, as c-FLIP, BAR, or cytosolic Bcl-2, and shortened by TRAIL receptors, caspase-8, Bid, and Bax. Moreover, the robustness of the switching time is determined on the level of Bax–Bcl-2 interaction leading to mitochondria pore formation (Albeck et al. 2008a). Thus, most modeling studies characterizing the temporal dynamics of apoptosis initiation in type II cells employing the mitochondrial pathway focused on regulation at caspase-8 or Bcl-2 level. In the following we will review the systems biological literature on the temporal dynamics of apoptosis initiation.

In the study of Hua et al. (2005) different topologies reflecting possible interactions of Bcl-2 with Bid, tBid, or Bax were compared regarding their role in controlling the kinetics of caspase-3 activation and preventing apoptosis. Four topologies, including Bcl-2 binding to Bid, tBid, or Bax only, or to tBid and Bax, were implemented into a large-scale model describing extrinsic apoptosis from ligand binding to caspase-3 activation. Specifically, their model describes DISC assembly, caspase-8 activation, Bid cleavage, and subsequent mitochondrial reactions (i.e., binding reactions of Bcl-2, Bax oligomerization, Smac release, cytochrome c release, and apoptosome formation). By comparing experimental data from wild type and Bcl-2 overexpressing cells with simulated trajectories of the model variants, model discrimination was possible: Their experimental data supported a mechanism where the caspase-3 time course reacts very sensitively to Bcl-2 overexpression, and the model suggested that this can only be realized if Bcl- 2 can simultaneously inhibit both tBid and Bax. A global sensitivity analysis on initial concentrations of the model variables gave insight into effects of overexpression or suppression. In this context it was interesting, that in the model of Hua et al. (2005) suppression of Bcl-2 under its basal level did not accelerate apoptosis, which could be experimentally verified. This indicates that Bcl-2 inhibitors could selectively sensitize Bcl-2 overexpressing tumor cells to apoptosis (Reed et al. 1996), while not affecting nontransformed cells expressing Bcl-2 at moderate levels. The sensitivity analysis by Hua et al. showed further asymmetric effects of overexpression or suppression: In some proteins an overexpression does not affect apoptosis timing, while their suppression causes significant changes (e.g., death receptors). Taken together, the model was able to predict dynamics of caspase-3 activation with high accuracy and provided insights into mechanisms of Bcl-2 action.

Albeck et al. (2008b) presented a refined model which exhibits similar complexity as the Hua et al. (2005) model, but takes into a large body of data, mostly at the single-cell level. Specifically, the model was trained against population data acquired with flow cytometry and western blotting as well as single cell imaging data. These were measured at various TRAIL concentrations and under several conditions of overexpression or depletion of signaling proteins. In the study of Albeck et al. (2008b) it was shown experimentally that MOMP timing and the kinetics of Smac/cytochrome c release were only dependent on the upstream signaling network controlling caspase-8 activation, while the contribution of positive feedback was negligible. Specifically, a role of three putative positive feedback mechanisms from caspase-3, caspase-6, or caspase-9 could be excluded. This suggests that some of the bistability mechanisms discussed above do not account for the temporal dynamics of apoptosis initiation at least in HeLa cells. In contrast to other apoptosis models the cell death decision was not dependent on bistability but explained apoptosis by the monostable transcritical process of MOMP. The interaction of Bcl-2 family members with tBid and Bax, the process of Smac release followed by the reduction of effector caspase inhibition by XIAP, and the mitochondria membrane pore formation dynamics were identified as critical stages of the all-or-none behavior of effector caspase activation. Signal transduction from MOMP to effector caspases apparently proceeds in a redundant manner through Smac inactivating XIAP and apoptosome formation causing XIAP sequestration and activation of caspase-3. In conclusion, the study by Albeck et al. currently represents the most comprehensive and realistic large-scale model of apoptosis.

In a combined experimental and theoretical study, Rehm et al. (2006) analyzed the kinetics of temporally switch-like effector caspase activation downstream of mitochondria. In particular, they focused on the control of effector caspase activation by XIAP. Their model described apoptosis signaling following MOMP induced by the drug staurosporine. The agent tetramethylrhodamine methylester (TMRM) was used to experimentally measure changes of the mitochondria membrane potential to monitor the occurrence of MOMP. In their model, Smac and cytochrome c released from mitochondria served as stimuli. Subsequent events in the model include apoptosome formation, caspase-9 as well as caspase-3 activation and caspase inhibition by XIAP. Their modeling analyses focused on the inhibition of caspase-9 and caspase-3 auto-amplification loops by XIAP, and model predictions were confirmed using single cell experiments with cells stably expressed FRET probes that contained a cleavage site for caspase-3. Specifically, their model prediction, that a reduction of XIAP levels would not affect apoptosis timing, while an XIAP overexpression would significantly delay effector caspase activation could be experimentally verified in HeLa cells and in MCF7 cells, which are completely devoid of caspase-3. Similarly, another prediction that interference on the level of Smac has only weak influences on apoptosis timing was verified. In conclusion the study of Rehm et al. (2006) provided detailed insights into the regulative function of XIAP on the timing of effector caspase activation.

We expected that future research will provide much more detailed insights into the kinetics of apoptosis initiation, since various tools are now available to monitor apoptosis on several signaling levels at the single-cell level: FRET reporters consisting of two fluorescent proteins linked with a cleavage sequence for caspase-8 or caspase-3 were used to monitor the activity of initiator and effector caspases (Albeck et al. 2008a; Rehm et al. 2002, 2006; Tyas et al. 2000). A GFP-tagged form of Bax was used to measure the translocation of Bax into mitochondria upon Bid activation (Albeck et al. 2008a). Furthermore in the study of Albeck et al. (2008a) a reporter protein that contains the mitochondria import sequence of Smac and thus accumulates in the mitochondria inter-membrane space (IMS-RP) was used to indicate MOMP in single cells. Simultaneous and quantitative description of multiple signaling levels will be a major challenge for future modeling studies.

1.5.4 Ambiguity of Cell Death or Survival

Experimental studies by Nair et al. (2004) had suggested that the mechanisms of cell death and survival can be mutually exclusive. The behavior of a population of cells upon an oxidative stress (H_2O_2) stimulus was observed. While some cells underwent apoptosis, others clearly showed a successful overcoming of the stress stimulus and a proliferative response. However, the paradigm of mutually exclusive cell death or survival processes was challenged in the subsequent experimental and theoretical studies described in this section.

To understand the double-edged role of CD95 (APO-1/Fas) activation in apoptosis as well as in NF-κB activation, models describing the role of c-FLIP on both cell fates were established (Fricker et al. 2010; Neumann et al. 2010), that are also described in chapter on extrinsic apoptosis. A focus of the models is the balance between caspase activation and inhibitory processes at the DISC. Cleavage of procaspase-8 homodimers bound to FADD can result in two forms that possess catalytic activity: an intermediate form p43 that remains bound to the DISC as a homodimer, and the completely processed form p18 that dissociates from the DISC as $p18_2$–$p10_2$ heterotetramers (Hoffmann et al. 2009; Hughes et al. 2009). Three splicing variants of the cellular FADD-like interleukin-1β-converting enzyme inhibitory protein (c-FLIP), c-FLIP short (c-$FLIP_S$), c-FLIP long ($FLIP_L$), and c-FLIP Raji ($FLIP_R$) can heterodimerize with a monomer of procaspase-8 bound to a FADD molecule at the DISC and interfere with caspase-8 activation. The two variants c-$FLIP_S$ and c-$FLIP_R$ block procaspase-8 autoprocessing in a heterodimer and therefore inhibit propagation of the apoptosis signal. In contrast, c-$FLIP_L$ can facilitate procaspase-8 cleavage to p43 but not to p18. Therefore, c-$FLIP_L$ leads to heterodimers with p43 at the DISC, and these complexes have a certain catalytic activity. While the two forms c-$FLIP_S$ and c-$FLIP_R$ clearly inhibit signal propagation, it was not obvious if c-$FLIP_L$ promotes or also inhibits apoptosis (Golks et al. 2005; Krueger et al. 2001; Scaffidi et al. 1999).

To resolve this question, a study of Fricker et al. (2010) on the signaling function of c-$FLIP_L$ combined experiments and modeling, and showed an ambiguous function of the protein as dependent on the stimulus strength. Their model considered the formation of homodimers of procaspase-8 or heterodimers of procaspase-8 and c-FLIP variants at the DISC and either termination of further reactions or processing to caspase-8 or p43-$FLIP_L$ by other active homo- or heterodimers. The model was trained against immunoblot data of procaspase-8 (p55), the intermediate form p43, and caspase-8 (p18 in $p18_2$–$p10_2$) at a given ligand concentration. The model predicted that at a low ligand concentration, a 20-fold overexpression of c-$FLIP_L$ would lead to a significant reduction of caspase-8 activity. Contrarily, it was predicted, that at high ligand concentration, the same overexpression would lead to an acceleration of cell death. The authors could confirm these predictions with time lapse imaging of cells moderately overexpressing c-$FLIP_L$. As the processing of procaspase-8 is relatively fast and c-$FLIP_L$ overexpression therefore can only cause small accelerating effects, Fricker et al. tested their model subsequently in conditions of c-$FLIP_S$ or c-$FLIP_R$ overexpression. As these c-FLIP variants inhibit caspase-8 activation, an even stronger activating effect of c-$FLIP_L$ was predicted. Again the predictions were verified, confirming the predicted ambiguous effect of c-$FLIP_L$ being antiapoptotic at low and proapoptotic at high stimulus strengths.

The study by Neumann et al. (2010) investigated in detail how NF-κB is activated by CD95L, and how this process is modulated by c-FLIP and caspase-8. The different possibilities of DISC formation of activated receptors, FADD, procaspase-8, c-FLIP$_L$, and c-FLIP$_S$ and their processed forms were described in their model (1) DISCs containing at least two procaspase-8 molecules give rise to p43 dimers and subsequently mature caspase-8. (2) At DISCs that contain at least one procaspase-8 and one c-FLIP$_L$ molecule the heterodimer p43-FLIP$_L$ is created. Besides, several other DISC constitutions preventing further processing were considered. By immunoprecipitation experiments, it was shown that complex formation of the heterodimer p43-FLIP$_L$ and IKKα leads to the phosphorylation of IκB (NF-κB·IκB·P). Phosphorylation of IκB was previously shown to trigger degradation, thereby promoting NF-κB activation. The subsequent translocation of p65 into the nucleus (NF-κB*) after IκB·P degradation was monitored in live cell experiments. The model matched the dynamics of these processes and the model complexity was iteratively reduced by summarizing or disregarding variables and modeled reactions. The simplification was supervised by means of the fitting accuracy in iterative parameter estimations. The balance between c-FLIP$_L$ and procaspase-8 was shown to be responsible for proliferative or apoptotic effects of CD95 stimulation. Regimes of predominant NF-κB activation or caspase-3 or of both processes could be predicted.

Systems biological studies on the ambiguity towards cell death or survival pathways revealed close interlinks between both processes. A more complicated overall image results when considering interlinks to other cellular signaling processes. Within the process of cell growth and mitosis, cell cycle regulatory mechanisms can determine pro- or anti-apoptotic conditioning as investigated in the cell cycle model of Pfeuty et al. (2008). Further interlinks exist between cell cycle repair mechanisms leading to cell cycle arrest or apoptosis execution (Zhang et al. 2009). Additionally, the apoptosis sensitivity is modulated by other cellular stress responses as theoretically analyzed in recent modeling studies (Bagci et al. 2008; Toivonen et al. 2011).

1.5.5 Understanding Cell-to-Cell Variability

Several new experimental techniques and theoretical approaches were developed to investigate the origins and the role of cellular variability in cell death signaling. In several modeling studies the effects of randomly choosing initial protein concentration sets from probability distributions to estimate consequences of variable initial protein concentrations were investigated (e.g., in Albeck et al. 2008b; Eissing et al. 2004; Toivonen et al. 2011; Würstle et al. 2010). This approach was employed by Eissing et al. to reconcile contradictions in single-cell and cell population-based measurements: Caspase activation is temporally abrupt at the single cell level (see above), while it apparently takes several hours when assessed by Western Blot measurements summing up over millions of cells. Using in silico simulations,

Eissing et al. could demonstrate that cell-to-cell variability in the lag time before caspase activation could explain the slow and gradual Western Blot time course. Thus, owing to cell-to-cell variability the fast kinetics of caspase-3 activation in single cells (Rehm et al. 2002) is masked in experimental observations of population data that show a gradual increase.

In general, it is not clear whether cell-to-cell variability in cell signaling mainly arises from stochastic dynamics in biochemical reactions (intrinsic noise) or from the variability of initial protein levels termed (extrinsic noise). Combining modeling and single cell experiments, it could be shown that heterogeneous apoptosis timing in a population of cells has its origin in the variable initial levels of apoptotic proteins (Spencer et al. 2009). In single cell experiments using FRET reporters for initiator and effector caspase activity and the MOMP reporter protein IMS-RP, after cell divisions pairs of cells were observed. While the time of apoptosis was highly correlated in the first hours after cell division, correlation in cell death timing ceased in pairs of older sister cells. This correlation was sustained over a longer period of time by inhibition of protein synthesis. Thereby, it was demonstrated that the variability of protein levels arising from noise in gene expression is responsible for cell-to-cell variability in apoptosis timing rather than genetic mutations or epigenetic differences that occur on a larger time scale. By combining single cell experiments with the model of Albeck et al. (2008b), it was verified that the time of apoptosis is dependent on the concentrations of several signaling proteins upstream of MOMP. Thus, the control of apoptosis timing appears to be distributed over many protein expression levels. Only when overexpressing single signaling proteins as Bid, the dependency of single protein concentration increases.

As an alternative approach to link the possible behavior of heterogeneous single cells to their population level measures at certain time points of a dynamic process, Hasenauer et al. (2011) developed methods based on parameter probability densities. These were demonstrated on a synthetic data set for a simple model of tumor necrosis factor (TNF) signaling (Chaves et al. 2008). The original parameter densities could successfully be estimated from artificial population snap shot data that represent flow cytometric data.

Cell-to-cell variability is an inherently complex phenomenon that can only be tackled by quantitative approaches. We therefore expect that ODE modeling combined with sensitivity analysis and stochastic modeling will be valuable tools in this field of apoptosis research.

1.6 Conclusions

Apoptosis is a well-characterized biological process amenable to mathematical modeling. Mechanistic models provided valuable insights into nonlinear phenomena such as all-or-none switching and irreversible decision making. Modeling reconciled apparently contradictory observations at the single-cell and population level, and was employed to identify molecular mechanisms controlling whether a cell enters death via the type I or type II pathway of apoptosis (Harrington et al. 2008; Hua et al. 2005; Okazaki et al. 2008). In particular, the study by Hua et al. (2005) verified corresponding model predictions experimentally, and showed that the antiapoptotic influence of Bcl-2 is completely lost in type II cells in the case of procaspase-8 overexpression. Nonintuitive feedback loops arising implicitly from the apoptosis network structure could be identified by simulation analysis (Cui et al. 2008; Legewie et al. 2006).

Less progress has been made concerning crosstalk between cell death and survival signaling; here, modeling was mostly restricted to top-down and qualitative approaches owing to crosstalk complexity. However, quantitative modeling and sensitivity analysis are required to predict effective cotreatment strategies for cancer cells that often harbor combined mutations in interdependent growth factor and apoptosis networks. Currently, a major limitation is our incomplete understanding of autophagy ("self eating") which shares many molecular components with the apoptosis machinery (e.g., Bcl-2 family members). Depending on the cell type, autophagy protects cells from death by removing damaged organelles or it triggers another form of cell death, further complicating the apoptosis survival network.

Another limitation of current models is that they mainly focus on extrinsic apoptosis induced by death ligands. However, most pharmacologically relevant responses, e.g., during chemotherapy, proceed via the intrinsic mitochondrial pathway. The intrinsic pathway includes another layer of complexity, as it requires transcriptional regulation, e.g., of Bcl-2 protein family regulators (PUMA, NOXA). Quantitative gene expression profiling and chromatin immunoprecipitation (ChIP) studies combined with systematic molecular perturbations are required to quantitatively model gene regulatory networks controlling intrinsic apoptosis. Antagonists inhibiting the Inhibitor of Apoptosis (IAPs) appear to be promising therapeutics, as they selectively kill cancer cells in the absence of further stimulation (Schimmer 2004); apoptosis models taking into account basal state signaling are required to understand and to optimize such therapeutic approaches.

Live-cell imaging and flow cytometric approaches led to insights into apoptosis at the single cell level, and revealed principles of cell-to-cell variability (Rehm et al. 2006; Spencer et al. 2009). Further experimental and theoretical analyzes are required to understand how complete eradication of tumor cell populations can be achieved. In principle, it might be possible that nonlinear phenomena such as bistability give rise to tumor cell sub-populations that are completely insensitive to therapy. Moreover, single cells may differ in the apoptosis pathways they employ, implying that combinatorial inhibition of multiple pathways is required for elimination of the whole tumor. This requires the development of new parameter estimation tools which take into account cell-to-cell variability, and integrate population-based as well as single-cell measurements.

References

Albeck JG, Burke JM, Aldridge BB et al (2008a) Quantitative analysis of pathways controlling extrinsic apoptosis in single cells. Mol Cell 30:11–25. doi:10.1016/j.molcel.2008.02.012

Albeck JG, Burke JM, Spencer SL et al (2008b) Modeling a snap-action, variable-delay switch controlling extrinsic cell death. PLoS Biol 6:2831–2852. doi:10.1371/journal.pbio.0060299

Aldridge BB, Haller G, Sorger PK, Lauffenburger DA (2006) Direct Lyapunov exponent analysis enables parametric study of transient signalling governing cell behaviour. Syst Biol 153:425–432

Bagci EZ, Vodovotz Y, Billiar TR et al (2006) Bistability in apoptosis: roles of bax, bcl-2, and mitochondrial permeability transition pores. Biophys J 90:1546–1559. doi:10.1529/biophysj.105.068122

Bagci EZ, Vodovotz Y, Billiar TR et al (2008) Computational insights on the competing effects of nitric oxide in regulating apoptosis. PLoS One 3:e2249. doi:10.1371/journal.pone.0002249

Bentele M, Lavrik I, Ulrich M et al (2004) Mathematical modeling reveals threshold mechanism in CD95-induced apoptosis. J Cell Biol 166:839–851. doi:10.1083/jcb.200404158

Breiman L, Friedman J, Stone CJ, Olshen RA (1984) Classification and regression trees (new edition). CRC, Boca Raton

Brunelle JK, Letai A (2009) Control of mitochondrial apoptosis by the Bcl-2 family. J Cell Sci 122:437–441. doi:10.1242/jcs.031682

Calzone L, Tournier L, Fourquet S et al (2010) Mathematical modelling of cell-fate decision in response to death receptor engagement. PLoS Comput Biol 6:e1000702. doi:10.1371/journal.pcbi.1000702

Certo M, Del Gaizo MV, Nishino M et al (2006) Mitochondria primed by death signals determine cellular addiction to antiapoptotic BCL-2 family members. Cancer Cell 9:351–365. doi:10.1016/j.ccr.2006.03.027

Chaves M, Eissing T, Allgöwer F (2008) Bistable biological systems: a characterization through local compact input-to-state stability. IEEE Trans Automat Contr 53:87–100

Chen C, Cui J, Lu H et al (2007a) Modeling of the role of a Bax-activation switch in the mitochondrial apoptosis decision. Biophys J 92:4304–4315. doi:10.1529/biophysj.106.099606

Chen C, Cui J, Zhang W, Shen P (2007b) Robustness analysis identifies the plausible model of the Bcl-2 apoptotic switch. FEBS Lett 581:5143–5150. doi:10.1016/j.febslet.2007.09.063

Chonghaile TN, Letai A (2008) Mimicking the BH3 domain to kill cancer cells. Oncogene 27(Suppl 1):S149–S157. doi:10.1038/onc.2009.52

Cui J, Chen C, Lu H et al (2008) Two independent positive feedbacks and bistability in the Bcl-2 apoptotic switch. PLoS One 3:e1469. doi:10.1371/journal.pone.0001469

Deng J, Carlson N, Takeyama K et al (2007) BH3 profiling identifies three distinct classes of apoptotic blocks to predict response to ABT-737 and conventional chemotherapeutic agents. Cancer Cell 12:171–185. doi:10.1016/j.ccr.2007.07.001

Dumitru CA, Gulbins E (2006) TRAIL activates acid sphingomyelinase via a redox mechanism and releases ceramide to trigger apoptosis. Oncogene 25:5612–5625. doi:10.1038/sj.onc.1209568

Düssmann H, Rehm M, Concannon CG et al (2010) Single-cell quantification of Bax activation and mathematical modelling suggest pore formation on minimal mitochondrial Bax accumulation. Cell Death Differ 17:278–290. doi:10.1038/cdd.2009.123

Eissing T, Conzelmann H, Gilles ED et al (2004) Bistability analyses of a caspase activation model for receptor-induced apoptosis. J Biol Chem 279:36892–36897. doi:10.1074/jbc.M404893200

Ferrell JE Jr (1996) Tripping the switch fantastic: how a protein kinase cascade can convert graded inputs into switch-like outputs. Trends Biochem Sci 21:460–466

Fischer SF, Vier J, Kirschnek S et al (2004) Chlamydia inhibit host cell apoptosis by degradation of proapoptotic BH3-only proteins. J Exp Med 200:905–916. doi:10.1084/jem.20040402

Fricker N, Beaudouin J, Richter P et al (2010) Model-based dissection of CD95 signaling dynamics reveals both a pro- and antiapoptotic role of c-FLIPL. J Cell Biol 190:377–389. doi:10.1083/jcb.201002060
Fussenegger M, Bailey JE, Varner J (2000) A mathematical model of caspase function in apoptosis. Nat Biotechnol 18:768–774. doi:10.1038/77589
Goldstein JC, Waterhouse NJ, Juin P et al (2000) The coordinate release of cytochrome c during apoptosis is rapid, complete and kinetically invariant. Nat Cell Biol 2:156–162. doi:10.1038/35004029
Golks A, Brenner D, Fritsch C et al (2005) c-FLIPR, a new regulator of death receptor-induced apoptosis. J Biol Chem 280:14507–14513. doi:10.1074/jbc.M414425200
Gómez-Benito M, Marzo I, Anel A, Naval J (2005) Farnesyltransferase inhibitor BMS-214662 induces apoptosis in myeloma cells through PUMA up-regulation, Bax and Bak activation, and Mcl-1 elimination. Mol Pharmacol 67:1991–1998. doi:10.1124/mol.104.007021
Harrington HA, Ho KL, Ghosh S, Tung KC (2008) Construction and analysis of a modular model of caspase activation in apoptosis. Theor Biol Med Model 5:26. doi:10.1186/1742-4682-5-26
Hasenauer J, Waldherr S, Doszczak M et al (2011) Identification of models of heterogeneous cell populations from population snapshot data. BMC Bioinformatics 12:125. doi:10.1186/1471-2105-12-125
Ho KL, Harrington HA (2010) Bistability in apoptosis by receptor clustering. PLoS Comput Biol 6:e1000956. doi:10.1371/journal.pcbi.1000956
Hoffmann JC, Pappa A, Krammer PH, Lavrik IN (2009) A new C-terminal cleavage product of procaspase-8, p30, defines an alternative pathway of procaspase-8 activation. Mol Cell Biol 29:4431–4440. doi:10.1128/MCB.02261-07
Hu T-M, Hayton WL, Mallery SR (2006) Kinetic modeling of nitric-oxide-associated reaction network. Pharm Res 23:1702–1711. doi:10.1007/s11095-006-9031-4
Hua F, Cornejo MG, Cardone MH et al (2005) Effects of Bcl-2 levels on Fas signaling-induced caspase-3 activation: molecular genetic tests of computational model predictions. J Immunol 175:985–995
Hua F, Hautaniemi S, Yokoo R, Lauffenburger DA (2006) Integrated mechanistic and data-driven modelling for multivariate analysis of signalling pathways. J R Soc Interface 3:515–526. doi:10.1098/rsif.2005.0109
Hughes MA, Harper N, Butterworth M et al (2009) Reconstitution of the death-inducing signaling complex reveals a substrate switch that determines CD95-mediated death or survival. Mol Cell 35:265–279. doi:10.1016/j.molcel.2009.06.012
Janes KA, Albeck JG, Gaudet S et al (2005) A systems model of signaling identifies a molecular basis set for cytokine-induced apoptosis. Science 310:1646–1653. doi:10.1126/science.1116598
Janes KA, Gaudet S, Albeck JG et al (2006) The response of human epithelial cells to TNF involves an inducible autocrine cascade. Cell 124:1225–1239. doi:10.1016/j.cell.2006.01.041
Krueger A, Schmitz I, Baumann S et al (2001) Cellular FLICE-inhibitory protein splice variants inhibit different steps of caspase-8 activation at the CD95 death-inducing signaling complex. J Biol Chem 276:20633–20640. doi:10.1074/jbc.M101780200
Legewie S, Blüthgen N, Herzel H (2006) Mathematical modeling identifies inhibitors of apoptosis as mediators of positive feedback and bistability. PLoS Comput Biol 2:e120. doi:10.1371/journal.pcbi.0020120
Levenberg K (1944) A method for the solution of certain problems in least squares. Quart Appl Math 2:164–168
Li B, Dou QP (2000) Bax degradation by the ubiquitin/proteasome-dependent pathway: involvement in tumor survival and progression. Proc Natl Acad Sci USA 97:3850–3855. doi:10.1073/pnas.070047997
Li P, Nijhawan D, Budihardjo I et al (1997) Cytochrome c and dATP-dependent formation of Apaf-1/caspase-9 complex initiates an apoptotic protease cascade. Cell 91:479–489
Mai Z, Liu H (2009) Boolean network-based analysis of the apoptosis network: irreversible apoptosis and stable surviving. J Theor Biol 259:760–769. doi:10.1016/j.jtbi.2009.04.024

Mannick JB, Hausladen A, Liu L et al (1999) Fas-induced caspase denitrosylation. Science 284:651–654

Marquardt D (1963) An algorithm for least-squares estimation of nonlinear parameters. SIAM J Appl Math 11:431–441

McDonald ER 3rd, El-Deiry WS (2004) Suppression of caspase-8- and -10-associated RING proteins results in sensitization to death ligands and inhibition of tumor cell growth. Proc Natl Acad Sci USA 101:6170–6175. doi:10.1073/pnas.0307459101

Nair VD, Yuen T, Olanow CW, Sealfon SC (2004) Early single cell bifurcation of pro- and antiapoptotic states during oxidative stress. J Biol Chem 279:27494–27501. doi:10.1074/jbc.M312135200

Neumann L, Pforr C, Beaudouin J et al (2010) Dynamics within the CD95 death-inducing signaling complex decide life and death of cells. Mol Syst Biol 6:352. doi:10.1038/msb.2010.6

Okazaki N, Asano R, Kinoshita T, Chuman H (2008) Simple computational models of type I/type II cells in Fas signaling-induced apoptosis. J Theor Biol 250:621–633. doi:10.1016/j.jtbi.2007.10.030

Pace V, Bellizzi D, Giordano F et al (2010) Experimental testing of a mathematical model relevant to the extrinsic pathway of apoptosis. Cell Stress Chaperones 15:13–23. doi:10.1007/s12192-009-0118-9

Pfeuty B, David-Pfeuty T, Kaneko K (2008) Underlying principles of cell fate determination during G1 phase of the mammalian cell cycle. Cell Cycle 7:3246–3257

Philippi N, Walter D, Schlatter R et al (2009) Modeling system states in liver cells: survival, apoptosis and their modifications in response to viral infection. BMC Syst Biol 3:97. doi:10.1186/1752-0509-3-97

Pop C, Fitzgerald P, Green DR, Salvesen GS (2007) Role of proteolysis in caspase-8 activation and stabilization. Biochemistry 46:4398–4407. doi:10.1021/bi602623b

Press WH, Vetterling WT, Teukolsky SA, Flannery BP (1992) Modeling of data, chap 15, 2nd edn, Numerical recipes in C. Cambridge University Press, New York, pp 662–666

Raue A, Kreutz C, Maiwald T et al (2009) Structural and practical identifiability analysis of partially observed dynamical models by exploiting the profile likelihood. Bioinformatics 25:1923–1929. doi:10.1093/bioinformatics/btp358

Raue A, Becker V, Klingmüller U, Timmer J (2010) Identifiability and observability analysis for experimental design in nonlinear dynamical models. Chaos 20:045105. doi:10.1063/1.3528102

Reed JC (1999) Dysregulation of apoptosis in cancer. J Clin Oncol 17:2941–2953

Reed JC, Miyashita T, Takayama S et al (1996) BCL-2 family proteins: regulators of cell death involved in the pathogenesis of cancer and resistance to therapy. J Cell Biochem 60:23–32. doi:10.1002/(SICI)1097-4644(19960101)60:1<23::AID-JCB5>3.0.CO;2-5

Rehm M, Dussmann H, Janicke RU et al (2002) Single-cell fluorescence resonance energy transfer analysis demonstrates that caspase activation during apoptosis is a rapid process. Role of caspase-3. J Biol Chem 277:24506–24514. doi:10.1074/jbc.M110789200

Rehm M, Huber HJ, Dussmann H, Prehn JHM (2006) Systems analysis of effector caspase activation and its control by X-linked inhibitor of apoptosis protein. EMBO J 25:4338–4349. doi:10.1038/sj.emboj.7601295

Rehm M, Huber HJ, Hellwig CT et al (2009) Dynamics of outer mitochondrial membrane permeabilization during apoptosis. Cell Death Differ 16:613–623. doi:10.1038/cdd.2008.187

Rössig L, Fichtlscherer B, Breitschopf K et al (1999) Nitric oxide inhibits caspase-3 by S-nitrosation in vivo. J Biol Chem 274:6823–6826

Scaffidi C, Schmitz I, Krammer PH, Peter ME (1999) The role of c-FLIP in modulation of CD95-induced apoptosis. J Biol Chem 274:1541–1548

Schimmer AD (2004) Inhibitor of apoptosis proteins: translating basic knowledge into clinical practice. Cancer Res 64:7183–7190. doi:10.1158/0008-5472.CAN-04-1918

Schlatter R, Schmich K, Avalos Vizcarra I et al (2009) ON/OFF and beyond–a boolean model of apoptosis. PLoS Comput Biol 5:e1000595. doi:10.1371/journal.pcbi.1000595

Schorr K, Li M, Krajewski S et al (1999) Bcl-2 gene family and related proteins in mammary gland involution and breast cancer. J Mammary Gland Biol Neoplasia 4:153–164

Scott FL, Stec B, Pop C et al (2009) The Fas-FADD death domain complex structure unravels signalling by receptor clustering. Nature 457:1019–1022. doi:10.1038/nature07606

Siehs C, Oberbauer R, Mayer G et al (2002) Discrete simulation of regulatory homo- and heterodimerization in the apoptosis effector phase. Bioinformatics 18:67–76

Song Z, Yao X, Wu M (2003) Direct interaction between survivin and Smac/DIABLO is essential for the anti-apoptotic activity of survivin during taxol-induced apoptosis. J Biol Chem 278:23130–23140. doi:10.1074/jbc.M300957200

Spencer SL, Gaudet S, Albeck JG et al (2009) Non-genetic origins of cell-to-cell variability in TRAIL-induced apoptosis. Nature 459:428–432. doi:10.1038/nature08012

Stennicke HR, Jürgensmeier JM, Shin H et al (1998) Pro-caspase-3 is a major physiologic target of caspase-8. J Biol Chem 273:27084–27090

Stucki JW, Simon H-U (2005) Mathematical modeling of the regulation of caspase-3 activation and degradation. J Theor Biol 234:123–131. doi:10.1016/j.jtbi.2004.11.011

Svingen PA, Loegering D, Rodriquez J et al (2004) Components of the cell death machine and drug sensitivity of the National Cancer Institute Cell Line Panel. Clin Cancer Res 10:6807–6820. doi:10.1158/1078-0432.CCR-0778-02

Timmer J, Müller T, Swameye I et al (2004) Modeling the nonlinear dynamics of cellular signal transduction. Int J Bifurcat Chaos 14:2069–2079

Toivonen HT, Meinander A, Asaoka T et al (2011) Modeling reveals that dynamic regulation of c-FLIP levels determines cell-to-cell distribution of CD95-mediated apoptosis. J Biol Chem 286:18375–18382. doi:10.1074/jbc.M110.177097

Tyas L, Brophy VA, Pope A et al (2000) Rapid caspase-3 activation during apoptosis revealed using fluorescence-resonance energy transfer. EMBO Rep 1:266–270. doi:10.1093/embo-reports/kvd050

Vieira HL, Belzacq AS, Haouzi D et al (2001) The adenine nucleotide translocator: a target of nitric oxide, peroxynitrite, and 4-hydroxynonenal. Oncogene 20:4305–4316

Vo T-T, Letai A (2010) BH3-only proteins and their effects on cancer. Adv Exp Med Biol 687:49–63

Willis SN, Chen L, Dewson G et al (2005) Proapoptotic Bak is sequestered by Mcl-1 and Bcl-xL, but not Bcl-2, until displaced by BH3-only proteins. Genes Dev 19:1294–1305. doi:10.1101/gad.1304105

Würstle ML, Laussmann MA, Rehm M (2010) The caspase-8 dimerization/dissociation balance is a highly potent regulator of caspase-8, -3, -6 signaling. J Biol Chem 285:33209–33218. doi:10.1074/jbc.M110.113860

Zhang H, Xu Q, Krajewski S et al (2000) BAR: an apoptosis regulator at the intersection of caspases and Bcl-2 family proteins. Proc Natl Acad Sci USA 97:2597–2602

Zhang R, Shah MV, Yang J et al (2008) Network model of survival signaling in large granular lymphocyte leukemia. Proc Natl Acad Sci USA 105:16308–16313. doi:10.1073/pnas.0806447105

Zhang T, Brazhnik P, Tyson JJ (2009) Computational analysis of dynamical responses to the intrinsic pathway of programmed cell death. Biophys J 97:415–434. doi:10.1016/j.bpj.2009.04.053

Chapter 2
Systems Biology of Death Receptor-Induced Apoptosis

Kolja Schleich and Inna N. Lavrik

Abstract Programmed cell death, termed apoptosis, plays a fundamental role in the development and homeostasis of multicellular organisms. Dysregulation of apoptosis can lead to numerous diseases, including autoimmune diseases, neurodegenerative diseases, and cancer. In mammalian cells apoptosis can be induced by intra- or extracellular stimuli. Extracellular stimuli comprise death ligands which lead to death receptor-induced apoptosis, referred to as extrinsic pathway. Intracellular signals, such as DNA damage, trigger the intrinsic pathway which results in the activation of Bcl-2 proteins and release of proapoptotic factors from the mitochondria into the cytosol. Apoptosis is executed by a family of cysteine proteases, the caspases, which eventually lead to the apoptotic phenotype, such as chromatin condensation, nuclear fragmentation, membrane blebbing, cell shrinkage, and formation of apoptotic bodies. The focus of this chapter is on understanding the signaling complexity of the extrinsic apoptotic pathway using systems biology. We summarize the main signaling paradigms and the major models of the extrinsic pathway. The development of these models has elucidated new insights into the regulation of apoptosis.

K. Schleich
Division of Immunogenetics, German Cancer Research Center (DKFZ), 69120 Heidelberg, Germany

Bioquant, 69120 Heidelberg, Germany

I.N. Lavrik (✉)
Department of Translational Inflammation Research, Institute of Experimental Internal Medicine, Otto von Guericke University, Leipziger Str. 44, 39120, Magdeburg, Germany
e-mail: inna.lavrik@med.ovgu.de

I.N. Lavrik (ed.), *Systems Biology of Apoptosis*, DOI 10.1007/978-1-4614-4009-3_2,

2.1 Death Receptor-Mediated Apoptosis

Death receptors (DR) belong to the tumor necrosis factor (TNF) family of proteins and are characterized by extracellular cysteine rich domains (CRD) and an intracellular ~80 amino acid long motif, the death domain (DD) (Ashkenazi 1998; Krammer et al. 2007). The best characterized DRs are CD95 (also named Fas/APO-1), TNFR1, TRAIL receptor 1 (TRAIL-R1), and TRAIL-R2 (Ashkenazi 1998; Krammer 2000; Krammer et al. 2007). Other DRs include DR3 and DR6, EDA-R, and NGF-R (Ashkenazi 1998; Krammer et al. 2007; Lavrik et al. 2005a). Death ligands (DL) are assumed to be homotrimeric (Yan and Shi 2005) and exist in a membrane-bound or a soluble form.

The CD95-induced apoptotic pathway is one of the best-studied signaling pathways. The natural ligand of CD95, CD95L, is expressed on a variety of cells, such as cytotoxic T cells, as a type II membrane protein (Krammer 2000; Rathmell et al. 1995). Cleavage of the membrane-bound to the soluble form reduces its apoptosis-inducing potential by more than 1,000-fold (Schneider 1998). Stimulation of CD95 with its ligand or with agonistic anti-CD95 antibodies, such as anti-APO-1 (Trauth et al. 1989), triggers the oligomerization of CD95 (Fig. 2.1). This leads to the recruitment of Fas-associated death domain (FADD) through DD interactions, as well as procaspase-8, procaspase-10, and cellular FLICE inhibitory proteins (c-FLIPs) via N-terminal death effector domains (DED), and formation of the death-inducing signaling complex (DISC). In the DISC procaspase-8/10 are activated by dimerization and internal cleavage (Lavrik et al. 2005b), which is regulated by c-FLIP proteins. Recently, it has been reported that ubiquitylation plays an important role in caspase-8 activation (Jin et al. 2009). Recent structural analyses challenge the concept of a trimeric ligand binding to a trimeric receptor. Scott et al. (2009) reported a tetrameric conformation of the CD95/FADD complex. Wang et al. (2010) suggest that the minimal signaling unit of CD95L is at least hexameric and found an asymmetric conformation of 5–7 CD95 bound to 5 FADD molecules.

Triggering of CD95 has also been reported to induce nonapoptotic pathways, such as NF-κB, AKT, and ERK (Neumann et al. 2010; Peter et al. 2007). However, the detailed mechanisms of the induction of CD95-mediated nonapoptotic pathways are not elucidated yet.

Activated caspase-8 is released from the DISC and activates effector caspases-3 and -7, which cleave a variety of substrates, such as DFF45/ICAD, thus releasing the DNase DFF40/CAD. DFF40/CAD is responsible for the fragmentation of chromosomes (Strasser et al. 2009; Yan and Shi 2005). Additionally, caspase-8 cleaves the BH3-only Bcl-2 protein Bid. The C-terminal part, tBid, then translocates to the mitochondria resulting in mitochondrial outer membrane permeabilization (MOMP) and the release of proapoptotic factors into the cytosol, such as cytochrome c, Apaf-1, or endonuclease G (Li 1998; Yan and Shi 2005). Recently, it has been reported that caspase-8 cleaves Bid in a specific complex at the mitochondria, which involves cardiolipin (Gonzalvez et al. 2008; Schug et al. 2011). This results in the

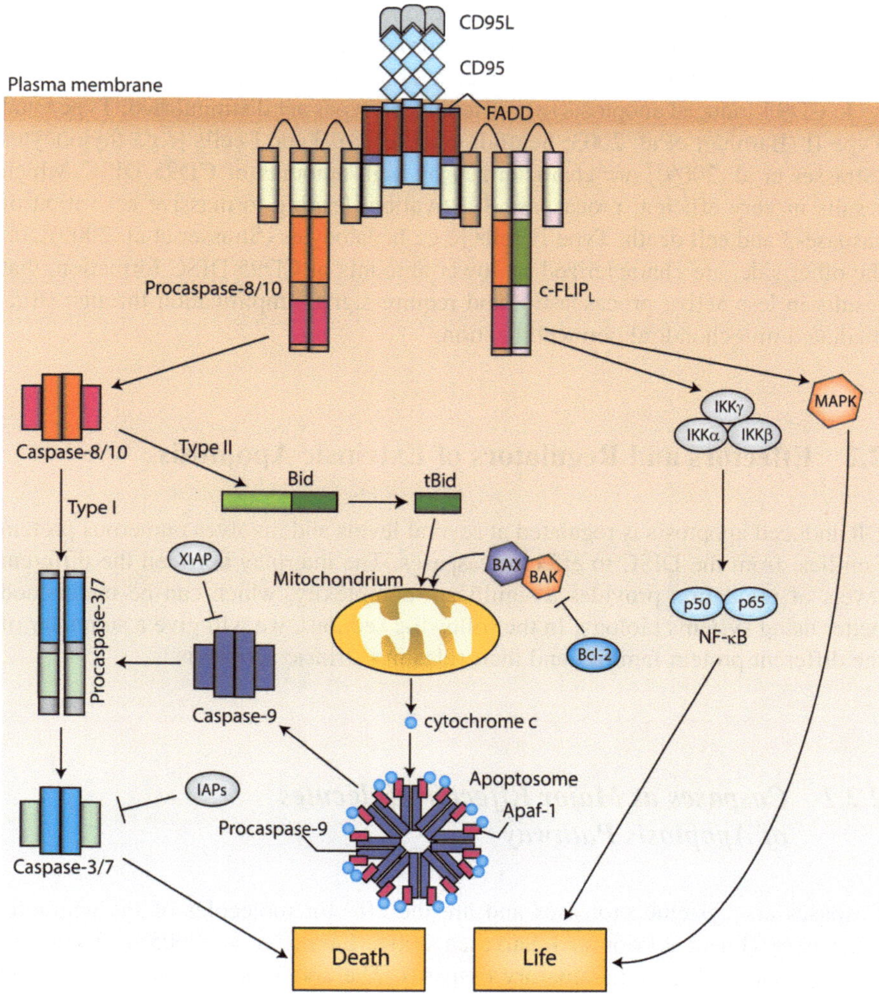

Fig. 2.1 Overview of CD95-induced pathways. Stimulation of CD95 leads to the formation of the death-inducing signaling complex (DISC), which includes at least CD95, FADD, procaspase-8, procaspase-10, and c-FLIP proteins. Caspase-8 is activated at the DISC which is regulated by c-FLIP proteins. In the cytosol caspase-8 cleaves and activates procaspase-3 or the Bcl-2 protein Bid. The truncated form of Bid (tBid) translocates to the mitochondria, which leads to mitochondrial outer membrane permeabilization (MOMP) and the release of cytochrome c, as well as other proapoptotic proteins from the mitochondria into the cytosol. Cytochrome c is involved in apoptosome formation and procaspase-9 activation. Procaspase-9 in turn also activates procaspase-3 resulting in caspase-3 activity and eventually cell death. Additionally CD95 stimulation can lead to the activation of nonapoptotic pathways, such as NF-κB or MAPK and cell survival

formation of another complex, the apoptosome, including cytochrome c, Apaf-1, ATP and procaspase-9, and activation of procaspase-9 in this complex (Boatright et al. 2003; Shi 2002). Caspase-9 also cleaves and activates caspase-3.

In CD95-induced apoptosis two different cell types are distinguished: Type I and Type II (Barnhart et al. 2003; Scaffidi et al. 1998). Type I cells [e.g., thymocytes (Strasser et al. 2009)] are characterized by high amounts of CD95 DISC, which results in very efficient procaspase-8 activation, leading to massive activation of caspase-3 and cell death. Type II cells [e.g., hepatocytes (Strasser et al. 2009)], on the other side, are characterized by lower amounts of CD95 DISC formation, that results in less active procaspase-8 and require signal amplification through tBid-mediated mitochondrial permeabilization.

2.2 Effectors and Regulators of Extrinsic Apoptosis

DR-induced apoptosis is regulated at several levels and involves numerous protein families, from the DISC to effector caspases. The interplay between the different levels of regulation provides a significant complexity, which can be understood better using systems biology. In the following sections, we will give a summary of the different protein families and their roles in extrinsic apoptosis.

2.2.1 Caspases as Major Effector Molecules of Apoptosis Pathway

Caspases are cysteine proteases and are the effector molecules of the apoptotic machinery (Fuentes-Prior and Salvesen 2004; Lavrik et al. 2005b). There are apoptotic, as well as inflammatory caspases. The apoptotic caspases are divided into initiator caspases, including caspase-2, -8, -9, and -10, and effector caspases, such as caspase-7 and -3 (Fuentes-Prior and Salvesen 2004). All caspases are present in the cell as inactive zymogens referred to as procaspases and are activated by internal cleavage (Fuentes-Prior and Salvesen 2004; Yan and Shi 2005). Initiator caspases act upstream of effector caspases and activate them through cleavage. Effector caspases then cleave a variety of cellular substrates, eventually resulting in cell death (Fuentes-Prior and Salvesen 2004). All caspases share a common structure. Caspase monomers consist of a large (~20 kDa) and small (~10 kDa) subunit (Fig. 2.2). Initiator caspases additionally have specific recruitment domains at their N-terminus. Procaspase-8 and -10 have two tandem DED through which they interact with FADD at the DISC, procaspase-9 possesses a caspase-recruitment domain (CARD) which is required for recruitment to the apoptosome via interactions with Apaf-1 (Fuentes-Prior and Salvesen 2004; Krammer et al. 2007; Lavrik et al. 2005b; Yan and Shi 2005).

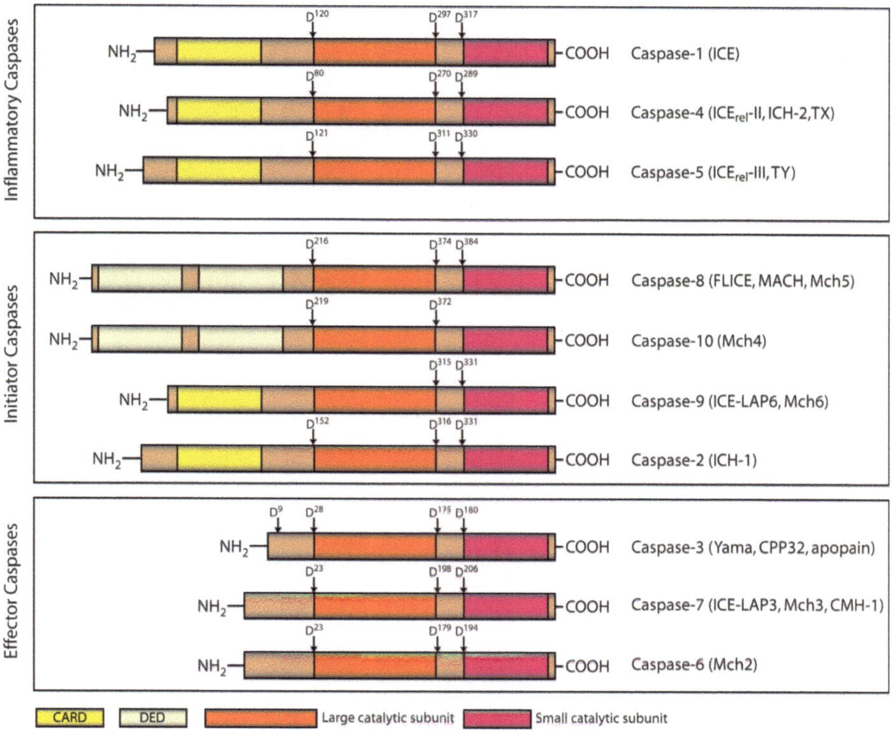

Fig. 2.2 Structural organization of caspases. Caspases are generally divided into inflammatory and apoptotic caspases. Apoptotic caspases are further divided into initiator and effector caspases. Caspases possess a large (p20) and small (p10) subunit. Initiator caspases additionally have DEDs (procaspase-8/10) or CARD domains (procaspase-9) at their N-terminus

Active caspases are heterotetramers consisting of two large and two small subunits (Fig. 2.3). Initiator caspases are present in the cytosol as monomers and are activated by dimerization or oligomerization at caspase-activating platforms and cleavage between the large and small subunits only stabilizes the dimer (Fuentes-Prior and Salvesen 2004; Hughes et al. 2009; Krammer et al. 2007; Lavrik et al. 2005b; Oberst et al. 2010). It was shown that both dimerization and interdomain cleavage are required for full activation of caspase-8 (Hughes et al. 2009; Keller et al. 2009; Oberst et al. 2010). Effector caspases on the contrary are present as inactive dimers and are readily activated by internal cleavage (Fuentes-Prior and Salvesen 2004; Oberst et al. 2010). This cleavage is performed by initiator caspases. The conceptual difference between the two classes is that there is no proteolytic enzyme upstream of initiator caspases. Consequently, initiator caspases exhibit low zymogenicity, which is defined as the ratio of activity between the cleaved and the uncleaved form. While initiator caspases have highest ratios of 10 (caspase-9) or 100 (caspase-8), the ratio for caspase-3 is more than 10,000 (Fuentes-Prior and Salvesen 2004).

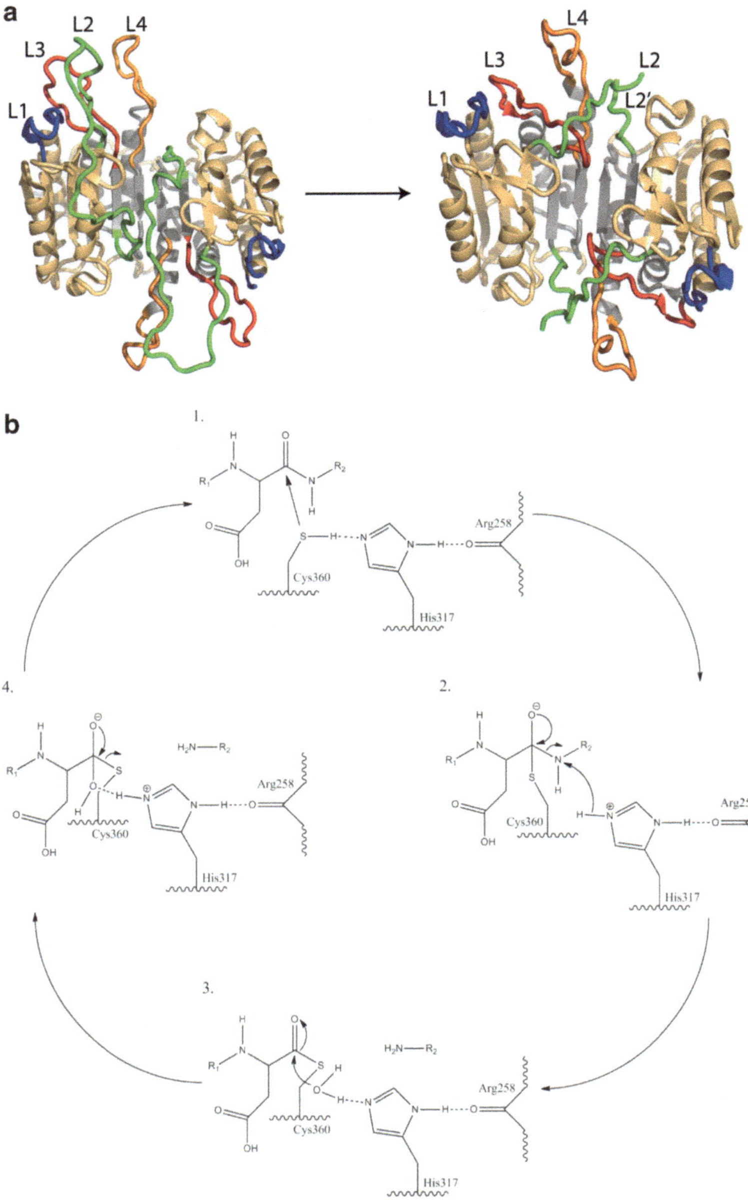

Fig. 2.3 Structure and suggested catalytic mechanism of caspases. (**a**) Crystal structure of procaspase-7 (*left*, PDB 1GQF) and active caspase-7 (*right*, PDB 1RHJ). The large and small subunits are colored *light orange* and *gray*, respectively. Cleavage of the intersubunit linker (L2) results in a conformational change in the active site and activation of caspase-7. The active site is formed by five loops, loops L1–L4 from one monomer and loop L2′ from the adjacent monomer. The structure was generated using PyMOL (Schrodinger, 2010). (**b**) The catalytic mechanism of caspases is suggested to be similar to serine proteases (Fuentes-Prior and Salvesen 2004). The cysteine acts as nucleophile and forms a tetrahedral intermediate with the substrate (step 2). A nearby histidine is critical for the subsequent hydrolysis of the substrate (step 3) [adapted from Fesik (2000)]

The active site of caspases is highly conserved and is formed by five protruding loops: L1, L2, L3, and L4 from one monomer and loop L2′ from the adjacent monomer (Fig. 2.3) (Fuentes-Prior and Salvesen 2004; Yan and Shi 2005). The catalytic cysteine residue is positioned in the loop L2 which determines, together with L1, L3, and L4 substrate specificity, recognizing specific tetrapeptide sequences. The L2′ loop from the adjacent monomer has mainly a stabilizing function of the active site (Fuentes-Prior and Salvesen 2004; Yan and Shi 2005). Initiator and effector caspases have a fundamental difference in active site architecture, which explains their different modes of activation. The intersubunit linker (L2) of initiator caspases is longer than that of effector caspases, which allows the formation of the active site also in its uncleaved form and results in the low zymogenicity (Fuentes-Prior and Salvesen 2004). The catalytic mechanism of caspases is suggested to be similar to serine proteases (Fuentes-Prior and Salvesen 2004). The cysteine acts as nucleophile and forms a tetrahedral intermediate with the substrate (Fig. 2.3b, step 2). A nearby histidine is critical for the subsequent hydrolysis of the substrate (Fig. 2.3b, step 3) (Fesik 2000; Fuentes-Prior and Salvesen 2004).

2.2.1.1 Caspase-8/10 Activation at the DISC

Two isoforms of procaspase-8 procaspase-8a (p55) and -8b (p53) are recruited to the DISC (Scaffidi et al. 1997). After binding to the DISC, procaspase-8a/b (p55/p53) undergoes processing, thus generating active caspase-8 (Lavrik et al. 2005b; Medema et al. 1997; Muzio et al. 1996) (Fig. 2.4). This processing has been suggested to occur via dimerization of two procaspase-8 monomers followed by a conformational change, leading to autoactivation of procaspase-8 homodimers (Chang et al. 2002; Golks et al. 2006b; Hughes et al. 2009; Yu et al. 2009). Procaspase-8a/b (p55/p53) processing at the DISC also results in the generation of the N-terminal cleavage products p43/p41, the prodomains p26/p24, as well as the C-terminal cleavage products p30, p18, and p10 (Golks et al. 2006b; Hoffmann et al. 2009; Hughes et al. 2009; Medema et al. 1997). Active caspase-8 heterotetramers $(p10/p18)_2$ generated at the DISC initiate the execution of apoptotis (Krammer et al. 2007). Recently, it has been reported that the cleavage products p30 and p43/p41 also possess catalytic activity leading to apoptosis initiation (Hoffmann et al. 2009; Hughes et al. 2009). Hence, procaspase-8 processing at the DISC initiates apoptosis through the generation of several catalytically active cleavage products. Recently it has been reported that full activation of caspase-8 at the DISC requires Cullin3-mediated polyubiquitination in the C-terminal part of caspase-8 (Jin et al. 2009).

Three isoforms of caspase-10 namely procaspase-10a, procaspase-10c, and procaspase-10d were reported to be bound to the DISC (Sprick et al. 2002). Procaspase-10 is also activated at the DISC via generation of homodimers, leading to the formation of an active heterotetramer. However, whether caspase-10 can

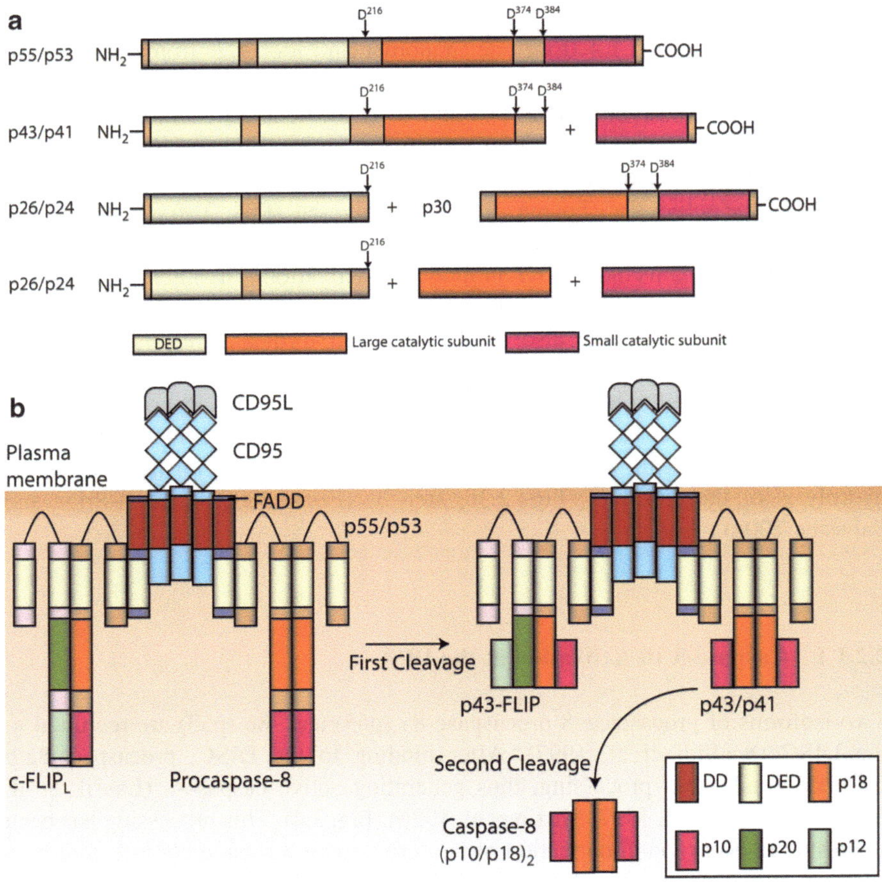

Fig. 2.4 Caspase-8 cleavage products and processing at the DISC. (**a**) Procaspase-8 can be cleaved between the large and small subunit or between the prodomain and the large catalytic subunit, resulting in numerous different cleavage products. (**b**) Procaspase-8 homodimers at the DISC can be processed to the active caspase-8 heterotetramers $(p10/p18)_2$ via the intermediate p43/p41. The procaspase-8/c-FLIP$_L$ heterodimer can only be processed to the intermediate p43-FLIP

trigger cell death in the absence of caspase-8 in response to CD95 stimulation is controversial (Fischer et al. 2006; Mühlethaler-Mottet et al. 2011; Sprick et al. 2002).

2.2.2 *c-FLIP Proteins Regulate Caspase Activation at the DISC*

Three c-FLIP isoforms and two cleavage products have been characterized so far (Golks et al. 2005, 2006a; Scaffidi et al. 1999; Ueffing et al. 2008) (Fig. 2.5). Three isoforms include one long, c-FLIP$_L$, and two short variants, c-FLIP$_S$ and c-FLIP$_R$. All three c-FLIP isoforms possess two tandem DED domains at their

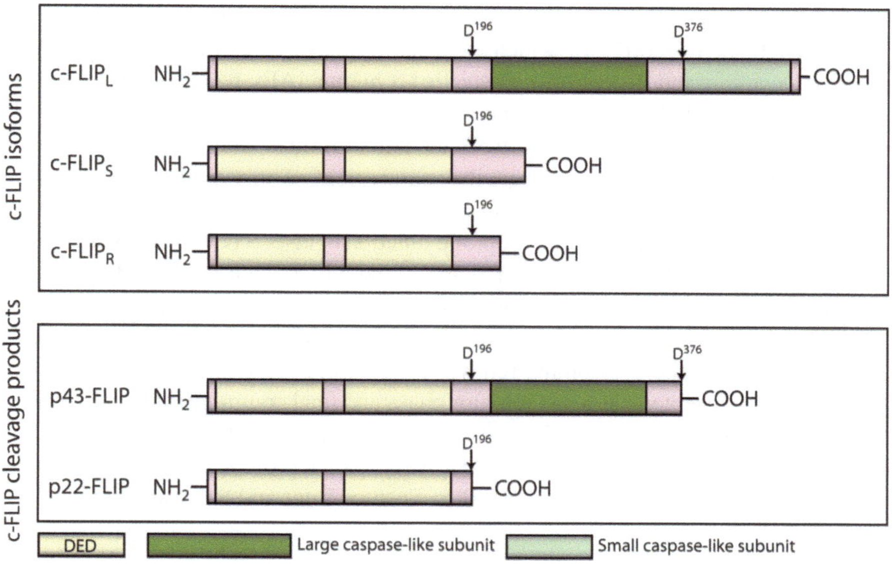

Fig. 2.5 c-FLIP isoforms and cleavage products. Three isoforms of c-FLIP proteins exist, one long (c-FLIP$_L$) and two short variants (c-FLIP$_S$ and c-FLIP$_R$). All isoforms contain two tandem DEDs which are required for DISC recruitment. c-FLIP$_L$ additionally has a large and small caspase-like subunit, which are catalytically inactive. C-FLIP$_L$ can be cleaved by caspase-8 at different positions generating the N-terminal fragment p43-FLIP or N-terminal fragment p22-FLIP

N-terminus. c-FLIP$_L$ additionally contains catalytically inactive caspase-like domains (p20 and p12). The two short isoforms, c-FLIP$_S$ and c-FLIP$_R$, block DR-induced apoptosis by inhibition of procaspase-8 activation at the DISC (Golks et al. 2005; Krueger et al. 2001). This has been suggested to occur through the formation of catalytically inactive procaspase-8/c-FLIP$_{R/S}$ heterodimers. c-FLIP$_L$ can play both a pro- and an antiapoptotic role. It can act as an antiapoptotic molecule, functioning in a way analogous to c-FLIP$_S$ when it is present at high concentrations at the DISC (Chang et al. 2002; Krueger et al. 2001). c-FLIP$_L$ can act proapoptotic when expressed at lower concentrations, in combination with strong receptor stimulation or in the presence of high amounts of either of the short c-FLIP isoforms c-FLIP$_S$ or c-FLIP$_R$. Under these conditions c-FLIP$_L$ facilitates the activation of procaspase-8 at the DISC (Fricker et al. 2010). This occurs via the formation of catalytically active procaspase-8/c-FLIP$_L$ heterodimers in which the procaspase-8 active loop is stabilized by c-FLIP$_L$ (Micheau et al. 2002; Yu et al. 2009), thereby increasing the catalytic activity of procaspase-8.

2.2.3 *IAP Family of Proteins*

The Inhibitors of Apoptosis (IAP) proteins directly inhibit caspases. They all share a conserved sequence motif of 70–80 amino acids, the baculoviral IAP repeat (BIR)

domain, which is arranged around a coordinated zinc atom (Fuentes-Prior and Salvesen 2004; Shi 2002), of which each family member can possess up to three copies (Deveraux and Reed 1999). There are six human IAPs, which include XIAP, c-IAP1, c-IAP2, NAIP, Bruce, and survivin (Deveraux and Reed 1999). Numerous mammalian IAPs, as well as IAPs in flies and viruses, possess a C-terminal RING domain, however, the requirement of this domain for apoptosis suppression remains unclear (Deveraux and Reed 1999). There are reports that both domains are required for their antiapoptotic function in insects (Clem and Miller 1994; Harvey et al. 1997), however, c-IAP1, c-IAP2, and XIAP in humans could still inhibit apoptosis when lacking the RING domain (Deveraux and Reed 1999; Hay et al. 1995).

XIAP directly inhibits active caspase-3. After MOMP Smac is released from the mitochondria into the cytosol and relieves XIAP-mediated inhibition. This causes a delay between receptor-mediated initiator phase and final commitment to cell death in type II cells (Albeck et al. 2008; Fuentes-Prior and Salvesen 2004). XIAP also contains a RING domain with E3 ubiquitin ligase activity, which promotes caspase-3 degradation by the proteasome (Albeck et al. 2008; Fuentes-Prior and Salvesen 2004).

2.2.4 Bcl-2 Family of Proteins

The Bcl-2 family of proteins play a key role in the regulation of apoptosis at the mitochondrial level and are essential for extrinsic apoptosis in type II cells or intrinsic apoptosis (Adams 1998; Yan and Shi 2005). The Bcl-2 protein family comprises at least 15 members with pro- as well as antiapoptotic functions (Adams 1998). All family members share a conserved structural motif, the Bcl-2 homology domain (BH1-BH4) (Adams 1998; Yan and Shi 2005). Prosurvival family members include Bcl-2 and Bcl-X_L. The proapoptotic Bcl-2 family members can be further subdivided into multidomain proteins, represented by Bax and Bak, and BH3-only proteins, such as Bid (Adams 1998; Yan and Shi 2005). The balance of pro- and antiapoptotic Bcl-2 family members determines the apoptosis induction. Proapoptotic Bcl-2 proteins cause the release of proapoptotic factors from the mitochondria by inducing MOMP. This process is triggered either in the extrinsic pathway through cleavage of Bid by caspase-8 or in the intrinsic pathway which is mainly controlled by the tumor suppressor p53 (Yu and Zhang 2005). The Bcl-2 family is described in detail in the Chaps. 4 and 5.

2.3 CD95-Induced Nonapoptotic Pathways

Accumulating evidence suggests that stimulation of CD95 can also induce nonapoptotic pathways, such as tumor growth and invasion, as well as proliferation and programmed necrosis, termed necroptosis (Chen et al. 2010; Choi et al. 2010; Geserick et al. 2009; Lee et al. 2011; Steller et al. 2011; Strasser et al. 2009; Tang et al. 2010). CD95-mediated nonapoptotic signaling occurs via induction of NF-κB, Akt and mitogen-activated protein kinases (MAPK) pathways. These, however, are not well understood, but have been reported to require caspase-8 activity (Kober et al. 2011; Lee et al. 2011; Nakajima et al. 2008; Neumann et al. 2010; Shikama et al. 2003; van Raam and Salvesen 2011). C-FLIP proteins play a very important role in the regulation of caspase-8 activation as well as induction of nonapoptotic pathways. It could be shown that the cleavage product of c-$FLIP_L$ p43-FLIP directly interacts with the IKK complex, leading to the activation of NF-κB (Neumann et al. 2010). Other prominent players in nonapoptotic pathways are receptor-interacting protein 1 (RIP1) and RIP3 which are important for CD95-induced necroptosis, as well as activation of NF-κB (Geserick et al. 2009; Kamarajan et al. 2010).

2.3.1 CD95-Mediated NF-κB Activation

The eukaryotic transcription factor NF-κB was originally discovered transcribing the immunoglobulin kappa light chain gene in B cells (Sen and Baltimore 1986). NF-κB can be activated following a variety of stimuli, including bacterial lipopolysaccharide (LPS), T cell receptor (TCR) signaling, different cytokines, such as TNFα, interleukin 1 (IL-1) and IL-2, viral infections, UV and X-ray radiation, nitric oxide, and also CD95 (Ghosh et al. 1998; Hayden and Ghosh 2004; Legembre et al. 2004; Verma et al. 1995). The NF-κB transcription family comprises multiple members, including RelA (p65), NF-κB1 (p50, p105), NF-κB2 (p52, p100), c-Rel, and RelB (Ghosh et al. 1998; Li and Verma 2002; Rasper et al. 1998; Verma et al. 1995). All NF-κB proteins share a conserved N-terminal 300-amino acid motif, the Rel homology domain (RHD), which contains a dimerization, nuclear localization as well as DNA-binding domain (Ghosh and Hayden 2008). NF-κB proteins form homo- or heterodimers with each other, except for RelB. The most prominent dimer which is commonly referred to as NF-κB is the heterodimer of p65 with either p50 or p52 (Li and Verma 2002). Importantly, only c-Rel, RelA, and RelB possess a transactivation domain and thus act as transcriptional activators, while other NF-κB proteins act as transcriptional repressors (Ghosh and Hayden 2008). In the absence of activating stimuli NF-κB dimers are inhibited by IκB (inhibitor of NF-κB) proteins via ankyrin-repeat motifs and masking their nuclear localization signal (Hayden et al. 2006). IκB proteins are phosphorylated by IκB kinases (IKKs) and subsequently degraded by the proteasome (Li and Verma 2002). The IKK complex, consisting of IKKα, IKKβ, and NEMO (IKKγ) regulates NF-κB activation following various stimuli and lack of either IKK complex component blocks NF-κB activation (Ghosh and Hayden 2008).

Besides its well studied proapoptotic function, CD95 also has nonapoptotic functions. It was shown that T cell proliferation under suboptimal CD3 stimulation could be enhanced by CD95 costimulation (Alderson et al. 1993; Paulsen and Janssen 2011; Paulsen et al. 2011). Additionally, CD95-knockout mice had abnormally low lymphocyte levels (Hao et al. 2004). Further, cancer cells were found to produce CD95L and inhibition of CD95 reduced tumor size in several cancer mouse models (Chen et al. 2010). The mechanism of the nonapoptotic functions of CD95 is largely unknown. It could be shown from different independent groups that CD95 stimulation triggers NF-κB activation, the exact mechanism, however, remains elusive (Kataoka et al. 2000; Kreuz et al. 2004; Neumann et al. 2010; O'Reilly et al. 2004). Numerous groups have reported that CD95-mediated activation of NF-κB, ERK, and MAPK requires caspase-8 activity and is controlled by c-FLIP proteins (Fricker et al. 2010; Golks et al. 2006a; Hughes et al. 2009; Kataoka et al. 2000; Kober et al. 2011; Kreuz et al. 2004; Nakajima et al. 2008; Neumann et al. 2010; Shikama et al. 2003; van Raam and Salvesen 2011). Our group could show in a systems biology study that p43-FLIP mediates NF-κB activation by direct interaction with the IKK complex (Neumann et al. 2010). However, the exact details of molecular mechanism of nonapoptotic signaling of CD95 will be addressed in further studies.

2.3.2 The Function of RIP in CD95-Mediated Nonapoptotic Pathways

The receptor-interacting protein (RIP) family plays a major role in CD95-mediated necrosis, as well as in life and death decisions after DR stimulation (Geserick et al. 2009, 2010; Kamarajan et al. 2010; Kreuz et al. 2004; Moquin and Chan 2010; Shikama et al. 2003; Vandenabeele et al. 2010a). The most prominent member is RIP1, a serine/threonine kinase, which also possesses one C-terminal DD, a RIP homotypic interaction motif (RHIM) and a ubiquitylation site (Geserick et al. 2009; Moquin and Chan 2010; Vandenabeele et al. 2010a, b). The DD of RIP1 allows it to interact with DR or FADD. Another important player is RIP3, a homolog of RIP1, lacking the DD. RIP1 and RIP3 can interact via their RHIM domains (Moquin and Chan 2010; Vandenabeele et al. 2010b). RIP1 can form another cytosolic complex, termed ripoptosome, which consists of at least RIP1, Caspase-8/10, and FADD and is regulated by cIAPs and c-FLIP proteins (Feoktistova et al. 2011; Tenev et al. 2011). The essential role of RIP1 for CD95-induced necroptosis, however, is controversial and it may be that both RIP1 and RIP3 are required (Bertrand and Vandenabeele 2010; Wong et al. 2010). While the kinase domain of RIP1 is required for the execution of necroptosis, ubiquitylation by cIAP1/2 has been reported to be needed for its nonapoptotic activities (Moquin and Chan 2010; Vandenabeele et al. 2010a, b). The exact roles of RIP1 and RIP3 in nonapoptotic pathways induced by CD95 are still unresolved and will be addressed in models of extrinsic apoptosis in the chapter by Zhivotovsky (Calzone et al. 2010). Especially the mechanistic interactions leading to nonapoptotic pathways after death receptor stimulation are poorly understood.

2.4 Mathematical Formalisms Used in Modeling of the Extrinsic Pathway

There are numerous different modeling formalisms to describe cellular systems and some of them have been applied for modeling the extrinsic pathway. The simplest model is a graph, which only qualitatively describes a system and cannot be used for simulation. Further, apoptosis signaling has been described by Boolean models (Calzone et al. 2010; Mai and Liu 2009; Saez-Rodriguez et al. 2009; Schlatter et al. 2009). The advantage of Boolean models is that simulation and optimization can be done very fast and no kinetic parameters need to be known, because each variable can be either on or off, represented by 1 or 0, respectively. The disadvantage, however, is that Boolean models are not quantitative and cannot represent reaction kinetics. Other modeling approaches which were also applied to apoptosis modeling include Bayesian models and petri nets (Heiner et al. 2004; Yang 2005). The most common approach used in apoptosis modeling is based on ordinary differential equations (ODEs), which describe cellular reactions based on mass-action kinetics and can describe biochemical networks with adequate accuracy. A first order ODE has the form $\dot{x} = f(t, x(t))$. Coupled differential equations can be used to describe the changes in protein concentrations depending on the system state. ODEs cannot describe diffusion in a cell. Diffusion is mostly neglected when modeling a signaling pathway, based on the general assumption that molecule diffusion is much faster compared to signal transduction. In order to include diffusion, partial differential equations (PDE) can be applied.

Contrary to models based on differential equations, which are deterministic, stochastic models use probabilities that two molecules interact with each other. This can be also used in agent-based models (Bonabeau 2002; Brown et al. 2011; Macal and North 2005, 2009; Roche et al. 2011). Each molecule is represented by an agent with certain properties based on biological knowledge. Agent-based models take into account local heterogeneity of molecules, while other approaches only consider molecule populations.

ODEs are constructed based on mass-action kinetics. The following chemical reaction:

$$A + B \underset{k_2}{\overset{k_1}{\rightleftharpoons}} C$$

transforms into the following system of ODEs:

$$\frac{\mathrm{d}[A]}{\mathrm{d}t} = \frac{\mathrm{d}[B]}{\mathrm{d}t} = -k_1 \cdot A \cdot B + k_2 \cdot C$$

$$\frac{\mathrm{d}[C]}{\mathrm{d}t} = k_1 \cdot A \cdot B - k_2 \cdot C$$

Usually ODEs cannot be solved analytically, but are solved numerically using computer programs, such as Matlab. The larger the model, the more kinetic parameters the system has. Measuring a large amount of parameters is often not possible resulting in unknown parameters. Either the system needs to be simplified to reduce parameter complexity or parameters need to be estimated. When estimating parameters the system should describe the system as accurately as possible. Parameter estimation can be done by the method of least-squares to fit experimental data to the model. Oftentimes the model includes many kinetic parameters compared to few experimental data, which gives multiple solutions for the parameter estimation. Therefore, sensitivity analysis is applied to test the robustness of the model upon parameter changes.

2.4.1 Modeling Extrinsic Apoptosis

The first model of CD95-mediated apoptosis was published more than a decade ago by Fussenegger et al. (2000), which was not based on experimental data, but only mechanistically described the apoptotic pathway. Albeck et al. (2008) quantitatively described death receptor-induced signaling with a focus on MOMP. Others focused on the intrinsic pathway and mitochondrial permeabilization (Düssmann et al. 2010; Rehm et al. 2009). The following section gives a review of systems biology studies of the CD95 signaling pathway as prototypic extrinsic pathway, all using an ODE-based modeling approach.

2.4.1.1 Revealing a Threshold Mechanism in CD95-Induced Apoptosis

In Bentele et al. (2004) the CD95 pathway topology was reconstructed by literature and database research. Initially, a model was generated consisting of about 70 molecules, 80 reactions, and more than 120 unknown parameters. The relatively high number of parameters did not allow parameter estimation using experimental data. Thus, subunits of varying information qualities were added. Well-known interactions were modeled mechanistically, and two “black boxes” were introduced for mitochondria and the degradation process. Black boxes were defined by their input–output behavior. For simplification the DISC was modeled as one complex, without taking into account its stoichiometry. These steps led to a model of CD95-induced apoptosis with 41 molecules (including complexes, e.g., DISC), 32 reactions, and two black boxes, namely mitochondria and degradation. The mitochondria black box takes the concentration of Bcl-X_L/Bcl-2 and tBid as input and triggers the release of cytochrome c if the concentration of tBid reaches a certain threshold compared to the concentrations of the antiapoptotic proteins. For degradation, a decay function was introduced, depending on the apoptotic activity of the different molecules. This model still had over 50 unknown parameters and was further simplified by applying sensitivity analysis, which tests the system stability upon parameter changes. Due to the large number of unknown parameters, the sensitivities for many randomly chosen points, covering a range of three orders of magnitude, were calculated. This analysis indicated that most sensitivities of the system were highly robust to changes in parameters. This analysis led to a simplified model of the DISC system in which the different c-FLIP isoforms were not distinguished and it was assumed that c-FLIP proteins block procaspase-8 activation at the DISC. Furthermore, the system dimensionality could be reduced from 58 to 18 unknown parameters. Based on this structured information model of CD95-induced cell death, experiments were designed to measure time-dependent concentrations of 15 different molecules after stimulation of CD95. The experiments were performed in the type I B lymphoblastoid cell line SKW6.4. The cells were stimulated with agonistic anti-APO-1 antibodies and the concentrations of the molecules of interest were measured by Western blot. In the first approach, kinetics of molecule concentrations were measured with an oversaturated ligand concentration of 5 μg/ml anti-APO-1, corresponding to a ligand/receptor ratio of about 5:1. The mathematical model could describe well the experimental data under these conditions. However, the system was still underdetermined, meaning different model parameters could match the same experimental data. Therefore, the same kinetics were measured for lower ligand concentrations and the parameters were estimated based on these different stimulation conditions. The resulting model could fit different stimulation scenarios. Most importantly, the model predicted a critical ligand concentration, below which apoptosis is abrogated, which could be verified experimentally (Bentele et al. 2004; Lavrik et al. 2007). It could be shown experimentally that c-FLIP proteins play a critical role in maintaining the threshold behavior of CD95 signaling. Upon low ligand concentration only few receptors would be activated and c-FLIP proteins could block all binding sites for caspase-8, thus preventing induction of apoptosis. High ligand concentrations, however, could lead to the activation of many receptors and caspase-8 would outnumber c-FLIP proteins at the DISC, allowing its activation and initiation of apoptosis.

2.4.1.2 The Stoichiometry of the DISC Determines Life and Death at CD95

In Neumann et al. (2010) life/death decisions after CD95 stimulation were analyzed. In particular, we focused on the crosstalk between the apoptotic and NF-κB signaling pathways. Initially a model was designed in which a trimerized ligand binds to trimerized CD95 further leading to the recruitment of three FADD molecules and DISC formation. To each of the FADD molecules either procaspase-8, c-FLIP$_L$, or c-FLIP$_S$ could be recruited via DED interactions. The order of binding of the DED proteins to the DISC gave rise to different intermediates. Fully formed DISCs were divided into three groups. The first group contained at least two procaspase-8 molecules leading to activation of procaspase-8 and apoptosis. The second group contained at least one procaspase-8 and one c-FLIP$_L$ molecule and would lead to the activation of NF-κB via generation of the c-FLIP$_L$ cleavage product p43-FLIP. The model assumed a direct interaction between p43-

FLIP and the IKK complex, leading to the phosphorylation and degradation of IκB and translocation of p65 to the nucleus. This prediction could be also verified experimentally. The third group of DISCs comprised all other conformations leading to a blocked state which neither initiates apoptosis nor NF-κB activation. The model was simulated using coupled ODEs and the large number of unknown parameters was reduced by assuming irreversible reactions. The remaining unknown parameters were estimated from experimental Western blot data by the method of least-squares. The model could well describe experimental data and postulated direct interaction between p43-FLIP and the IKK complex. Interestingly effector caspase-activity was not required to reproduce the dynamics of NF-κB activation after CD95 stimulation. In addition, the model postulated that apoptotic and nonapoptotic pathways diverge at the DISC. This hypothesis was tested experimentally by inhibiting the apoptotic branch using the pan-caspase inhibitor zVAD-fmk, which blocks caspase activation downstream of the DISC, but not at the DISC (Golks et al. 2006b; Hughes et al. 2009). We stimulated zVAD-fmk pretreated CD95 overexpressing HeLa cells (HeLa-CD95) and studied NF-κB activation as well as apoptotic signaling. We could see p65 translocation to the nucleus using mCherry-tagged p65 and fluorescence microscopy, but could not detect apoptosis, thus validating the model prediction. In order to study whether the complete model might hide key features due to its complexity we performed a stepwise model reduction. Parameter estimation was done after each reducing step and its performance compared to the original complete model. A simpler model usually gives more reliable parameter estimation, because very complex models are prone to overfitting. We came up with a simplified model, in which CD95 and FADD were assumed to be pretrimerized (RF). Upon ligand (L) binding, the L/RF complex could further recruit procaspase-8, c-FLIP$_L$, or c-FLIP$_S$. Two procaspase-8 molecules in the DISC would lead to the generation of active caspase-8 heterotetramers (p18/p10)$_2$ via p43/p41, while the presence of one molecule of procaspase-8 and c-FLIP$_L$ initiates the NF-κB pathway via the cleavage product p43-FLIP leading to nonlinear dynamics. Recruitment of c-FLIP$_S$ to the DISC leads to inhibition. This model could still well describe experimental data and also fit with the previously found threshold behavior of CD95-signaling (Bentele et al. 2004). Interestingly, the threshold concentration did not depend on the number of CD95 on the cell surface, but determines the rates of apoptotic and nonapoptotic signaling, which was found by comparing HeLa wt and HeLa-CD95 (Neumann et al. 2010). Importantly, we revealed that the stoichiometry of DISC components and especially the ratios of c-FLIP isoforms to procaspase-8 and the concentration of their cleavage products p43/p41 and p43-FLIP play a crucial role in the life/death decisions in CD95 signaling.

2.4.1.3 Dual Function of c-FLIP$_L$ in Procaspase-8 Processing and Cell Death

In Fricker et al. (2010) we also focused on the CD95 DISC and studied the role of c-FLIP proteins in the activation of procaspase-8 and cell death. We studied the influence of c-FLIP on caspase-8 activation in HeLa-CD95 cells which stably overexpress c-FLIP$_L$ (HeLa-CD95-F$_L$), c-FLIP$_S$ (HeLa-CD95-F$_S$), or c-FLIP$_R$ (HeLa-CD95-F$_R$). Procaspase-8 activation was studied after stimulation of these cell lines for different times with CD95L, followed by CD95 DISC Immunoprecipitation (IP) and Western blot analysis. Overexpression of either of the short isoforms of c-FLIP resulted in inhibition of procaspase-8 processing, even after several hours. Interestingly overexpression of c-FLIP$_L$ strongly accelerated the first cleavage step to p43/p41, which agreed with previous studies (Golks et al. 2005; Krueger et al. 2001; Micheau et al. 2002). This effect was also observed for different concentrations of CD95L. Even though the first cleavage of procaspase-8 was enhanced, c-FLIP$_L$ overexpressing HeLa-CD95 cells were less sensitive to apoptosis induction compared to normal HeLa-CD95. Stimulation with lower concentrations of CD95L resulted in complete inhibition of procaspase-8 processing as well as caspase-3 and PARP cleavage in c-FLIP$_L$ overexpressing HeLa-CD95 cells, but not in normal HeLa-CD95 cells. Stimulation with a high dose of ligand led to accelerated caspase-8 processing and cleavage of caspase-3 and PARP. Essentially, we observed nonlinear effects of c-FLIP$_L$ overexpression on procaspase-8 processing and cell death. To further study these effects, we built a mathematical model of caspase-8 processing at the DISC using coupled ODEs. The model involved CD95L which binds to CD95, further causing the recruitment of FADD to form the DISC. Contrary to the model of Neumann et al. (2010), we considered the complex of CD95L/CD95/FADD as monomer which can further recruit one DED protein. Procaspase-8 can form three different dimers at the DISC: procaspase-8 homodimers, procaspase-8/c-FLIP$_L$ heterodimers, and procaspase-8/c-FLIP$_{S/R}$ heterodimers. Procaspase-8 homodimers could be fully processed to p43/p41 and p18, procaspase-8/c-FLIP$_L$ heterodimers could be only processed to p43/p41, but not p18, and procaspase-8/c-FLIP$_{S/R}$ heterodimers could not be processed at all. This assumption was experimentally validated in HeLa-CD95-c-FLIP$_{L/R}$ cells, overexpressing both the short and long isoforms of c-FLIP. Furthermore, procaspase-8 homodimers and procaspase-8/c-FLIP$_L$ heterodimers were assumed to be catalytically active and cleave other molecules at the DISC. Fully processed caspase-8 heterotetramers would be released into the cytosol and replaced by new DED proteins, causing a turnover of caspase-8 at the DISC. Additionally, to reduce model complexity, procaspase-8a and procaspase-8b were considered as one entity, because they had been reported to have similar catalytic properties (Hughes et al. 2009). In addition, processing of procaspase-8 to p30 was neglected due to much lower amounts compared to p43/p41 (Hoffmann et al. 2009). Essentially, we showed that c-FLIP$_L$ only acts proapoptotic under certain conditions and depends upon the strength of stimulation. At moderate concentrations at the DISC combined with strong stimulation of CD95 or high amounts of one of the short c-FLIP isoforms at the DISC c-FLIP$_L$ plays a proapoptotic role, while high amounts of c-FLIP$_L$ at the DISC results in inhibition of apoptotic signaling. Due to its high affinity to the DISC (Chang et al. 2002), c-FLIP$_L$ could be preferentially recruited to the DISC and compete with procaspase-8 for binding sites. In addition c-FLIP$_L$ could block procaspase-8 turnover at the DISC by blocking the final cleavage step to p18.

Finally, the role of c-$FLIP_L$ in procaspase-8 processing is further modulated by the short c-FLIP isoforms. High levels of c-$FLIP_{R/S}$ in the cell cause a sensitizing effect of c-$FLIP_L$. These findings demonstrate that c-FLIP proteins are critical regulators of life and death decisions in CD95 signaling. Further, they demonstrate the importance of the stoichiometry of the DISC in this decision process. Changes in DISC components could shift the signaling from apoptotic to nonapoptotic or vice versa. In addition the data show the complex interplay of the different components at the DISC, because the action of c-$FLIP_L$ does not only depend on its own concentration but also on the concentration of the short variants at the DISC as well as the strength of stimulation.

2.5 Conclusions

CD95-induced apoptosis is one of the best-studied signaling pathways, making it especially interesting for modeling. Stimulation of CD95 can induce, both, apoptotic and nonapoptotic responses. The apoptotic response upon CD95 stimulation is very well defined. CD95 stimulation leads to the formation of the DISC and activation of the initiator procaspases-8 and -10 that, in turn, triggers the apoptotic cascade.

Apoptosis has been subject of intense modeling, using different modeling formalisms, including Boolean modeling (Calzone et al. 2010; Mai and Liu 2009; Saez-Rodriguez et al. 2009; Schlatter et al. 2009), Bayesian modeling, petri nets (Heiner et al. 2004; Yang 2005), and ODEs (Albeck et al. 2008; Bentele et al. 2004; Cui et al. 2008; Fricker et al. 2010; Fussenegger et al. 2000; Kober et al. 2011; Legewie et al. 2006; Neumann et al. 2010; Rehm et al. 2006; Spencer and Sorger 2011). We could successfully model extrinisic apoptosis using coupled ODEs based on biochemical data and derive reasonable biological conclusions. Using this systems biology approach, we could reveal the threshold mechanism of CD95-induced apoptosis (Bentele et al. 2004; Lavrik et al. 2007). Furthermore, we could show that the stoichiometry of the DISC is central to life and death decisions upon CD95-stimulation (Neumann et al. 2010). In the same study, we showed that p43-FLIP directly interacts with the IKK complex leading to NF-κB activation after CD95 stimulation. In addition, we could gain detailed insights into the role of c-FLIP proteins in the activation of apoptosis or nonapoptotic pathways and showed that c-$FLIP_L$ can exhibit pro- or antiapoptotic functions, depending on the strength of stimulation, as well as its concentration at the DISC (Fricker et al. 2010).

Our results depicted the important function of the DISC in the activation of apoptotic as well as nonapoptotic responses. Especially the DISC stoichiometry and the complex interplay between DISC components decide the signaling outcome. The stoichiometry of the DISC, however, has only been studied for CD95 and FADD (Scott et al. 2009; Wang et al. 2010), but no conclusive data on all major components, including caspase-8/10 as well as c-FLIP proteins are available.

Determining the stoichiometry of the CD95 DISC and comparing apoptotic vs. nonapoptotic cells could provide key insights into this decision process, which is still not understood conclusively. Analysis of the CD95 DISC stoichiometry using mathematical modeling should provide new insights into the regulation of CD95 signaling. Furthermore, it could be possible to alter the DISC stoichiometry pharmacologically, thus switching between the two signaling outcomes in CD95. This would add new therapies for diseases with dysfunctional apoptosis, such as neurodegenerative disorders or cancer.

Acknowledgments We acknowledge the Helmholtz Alliance on Systems Biology (NW1SB-Cancer) and Helmholtz-Russia Joint Research Groups-2008-2 for supporting our work.

References

Adams JM (1998) The Bcl-2 protein family: arbiters of cell survival. Science 281:1322–1326

Albeck JG, Burke JM, Aldridge BB, Zhang M, Lauffenburger DA, Sorger PK (2008) Quantitative analysis of pathways controlling extrinsic apoptosis in single cells. Mol Cell 30:11–25

Alderson MR, Armitage RJ, Maraskovsky E, Tough TW, Roux E, Schooley K, Ramsdell F, Lynch DH (1993) Fas transduces activation signals in normal human T lymphocytes. J Exp Med 178:2231–2235

Ashkenazi A (1998) Death receptors: signaling and modulation. Science 281:1305–1308

Barnhart BC, Alappat EC, Peter ME (2003) The CD95 Type I/Type II model. Semin Immunol 15:185–193

Bentele M, Lavrik I, Ulrich M, Stösser S, Heermann DW, Kalthoff H, Krammer PH, Eils R (2004) Mathematical modeling reveals threshold mechanism in CD95-induced apoptosis. J Cell Biol 166:839–851

Bertrand MJM, Vandenabeele P (2010) RIP1's function in NF-κB activation: from master actor to onlooker. Cell Death Differ 17:379–380

Boatright KM, Renatus M, Scott FL, Sperandio S, Shin H, Pedersen IM, Ricci JE, Edris WA, Sutherlin DP, Green DR, Salvesen GS (2003) A unified model for apical caspase activation. Mol Cell 11:529–541

Bonabeau E (2002) Agent-based modeling: methods and techniques for simulating human systems. Proc Natl Acad Sci USA 99(Suppl 3):7280–7287

Brown BN, Price IM, Toapanta FR, Dealmeida DR, Wiley CA, Ross TM, Oury TD, Vodovotz Y (2011) An agent-based model of inflammation and fibrosis following particulate exposure in the lung. Math Biosci 231(2):186–96

Calzone L, Tournier L, Fourquet S, Thieffry D, Zhivotovsky B, Barillot E, Zinovyev A (2010) Mathematical modeling of cell-fate decision in response to death receptor engagement. PLoS Comput Biol 6:e1000702

Chang DW, Xing Z, Pan Y, Algeciras-Schimnich A, Barnhart BC, Yaish-Ohad S, Peter ME, Yang X (2002) c-FLIP(L) is a dual function regulator for caspase-8 activation and CD95-mediated apoptosis. EMBO J 21:3704–3714

Chen L, Park S-M, Tumanov AV, Hau A, Sawada K, Feig C, Turner JR, Fu Y-X, Romero IL, Lengyel E, Peter ME (2010) CD95 promotes tumour growth. Nature 465:492–496

Choi K, Ni L, Jonakait GM (2010) Fas ligation and tumor necrosis factor alpha activation of murine astrocytes promote heat shock factor-1 activation and heat shock protein expression leading to chemokine induction and cell survival. J Neurochem 116(3):438–448

Clem RJ, Miller LK (1994) Control of programmed cell death by the baculovirus genes p35 and iap. Mol Cell Biol 14:5212–5222

Cui J, Chen C, Lu H, Sun T, Shen P (2008) Two independent positive feedbacks and bistability in the Bcl-2 apoptotic switch. PLoS One 3:e1469
Deveraux QL, Reed JC (1999) IAP family proteins–suppressors of apoptosis. Genes Dev 13:239–252
Düssmann H, Rehm M, Concannon CG, Anguissola S, Würstle M, Kacmar S, Völler P, Huber HJ, Prehn JHM (2010) Single-cell quantification of Bax activation and mathematical modeling suggest pore formation on minimal mitochondrial Bax accumulation. Cell Death Differ 17:278–290
Feoktistova M, Geserick P, Kellert B, Dimitrova DP, Langlais C, Hupe M, Cain K, Macfarlane M, Häcker G, Leverkus M (2011) cIAPs block ripoptosome formation, a RIP1/Caspase-8-containing intracellular cell death complex differentially regulated by cFLIP isoforms. Mol Cell 43(3–6):449–463
Fesik SW (2000) Insights into programmed cell death through structural biology. Cell 103:273–282
Fischer U, Stroh C, Schulze-Osthoff K (2006) Unique and overlapping substrate specificities of caspase-8 and caspase-10. Oncogene 25:152–159
Fricker N, Beaudouin J, Richter P, Eils R, Krammer PH, Lavrik IN (2010) Model-based dissection of CD95 signaling dynamics reveals both a pro- and antiapoptotic role of c-FLIPL. J Cell Biol 190:377–389
Fuentes-Prior P, Salvesen GS (2004) The protein structures that shape caspase activity, specificity, activation and inhibition. Biochem J 384:201–232
Fussenegger M, Bailey JE, Varner J (2000) A mathematical model of caspase function in apoptosis. Nat Biotechnol 18:768–774
Geserick P, Hupe M, Moulin M, Wong WW-L, Feoktistova M, Kellert B, Gollnick H, Silke J, Leverkus M (2009) Cellular IAPs inhibit a cryptic CD95-induced cell death by limiting RIP1 kinase recruitment. J Cell Biol 187:1037–1054
Geserick P, Hupe M, Moulin M, Leverkus M (2010) RIP-in CD95-induced cell death: the control of alternative death receptors pathways by cIAPs. Cell Cycle 9:2689–2691
Ghosh S, Hayden MS (2008) New regulators of NF-κB in inflammation. Nat Rev Immunol 8:837–848
Ghosh S, May MJ, Kopp EB (1998) NF-κ B and Rel proteins: evolutionarily conserved mediators of immune responses. Annu Rev Immunol 16:225–260
Golks A, Brenner D, Fritsch C, Krammer PH, Lavrik IN (2005) c-FLIPR, a new regulator of death receptor-induced apoptosis. J Biol Chem 280:14507–14513
Golks A, Brenner D, Krammer PH, Lavrik IN (2006a) The c-FLIP-NH2 terminus (p22-FLIP) induces NF-κB activation. J Exp Med 203:1295–1305
Golks A, Brenner D, Schmitz I, Watzl C, Krueger A, Krammer PH, Lavrik IN (2006b) The role of CAP3 in CD95 signaling: new insights into the mechanism of procaspase-8 activation. Cell Death Differ 13:489–498
Gonzalvez F, Schug ZT, Houtkooper RH, MacKenzie ED, Brooks DG, Wanders RJA, Petit PX, Vaz FM, Gottlieb E (2008) Cardiolipin provides an essential activating platform for caspase-8 on mitochondria. J Cell Biol 183:681–696
Hao Z, Hampel B, Yagita H, Rajewsky K (2004) T cell-specific ablation of Fas leads to Fas ligand-mediated lymphocyte depletion and inflammatory pulmonary fibrosis. J Exp Med 199:1355–1365
Harvey AJ, Soliman H, Kaiser WJ, Miller LK (1997) Anti- and pro-apoptotic activities of baculovirus and Drosophila IAPs in an insect cell line. Cell Death Differ 4:733–744
Hay BA, Wassarman DA, Rubin GM (1995) Drosophila homologs of baculovirus inhibitor of apoptosis proteins function to block cell death. Cell 83:1253–1262
Hayden MS, Ghosh S (2004) Signaling to NF-κB. Genes Dev 18:2195–2224
Hayden MS, West AP, Ghosh S (2006) NF-κB and the immune response. Oncogene 25:6758–6780
Heiner M, Koch I, Will J (2004) Model validation of biological pathways using Petri nets–demonstrated for apoptosis. Biosystems 75:15–28

Hoffmann JC, Pappa A, Krammer PH, Lavrik IN (2009) A new C-terminal cleavage product of procaspase-8, p30, defines an alternative pathway of procaspase-8 activation. Mol Cell Biol 29:4431–4440

Hughes MA, Harper N, Butterworth M, Cain K, Cohen GM, MacFarlane M (2009) Reconstitution of the death-inducing signaling complex reveals a substrate switch that determines CD95-mediated death or survival. Mol Cell 35:265–279

Jin Z, Li Y, Pitti R, Lawrence D, Pham VC, Lill JR, Ashkenazi A (2009) Cullin3-based polyubiquitination and p62-dependent aggregation of caspase-8 mediate extrinsic apoptosis signaling. Cell 137:721–735

Kamarajan P, Bunek J, Lin Y, Nunez G, Kapila YL (2010) Receptor-interacting protein shuttles between cell death and survival signaling pathways. Mol Biol Cell 21:481–488

Kataoka T, Budd RC, Holler N, Thome M, Martinon F, Irmler M, Burns K, Hahne M, Kennedy N, Kovacsovics M, Tschopp J (2000) The caspase-8 inhibitor FLIP promotes activation of NF-κB and Erk signaling pathways. Curr Biol 10:640–648

Keller N, Mares J, Zerbe O, Grütter MG (2009) Structural and biochemical studies on procaspase-8: new insights on initiator caspase activation. Structure 17:438–448

Kober AMM, Legewie S, Pforr C, Fricker N, Eils R, Krammer PH, Lavrik IN (2011) Caspase-8 activity has an essential role in CD95/Fas-mediated MAPK activation. Cell Death Dis 2:e212

Krammer PH (2000) CD95's deadly mission in the immune system. Nature 407:789–795

Krammer PH, Arnold R, Lavrik IN (2007) Life and death in peripheral T cells. Nat Rev Immunol 7:532–542

Kreuz S, Siegmund D, Rumpf J-J, Samel D, Leverkus M, Janssen O, Häcker G, Dittrich-Breiholz O, Kracht M, Scheurich P, Wajant H (2004) NFkappaB activation by Fas is mediated through FADD, caspase-8, and RIP and is inhibited by FLIP. J Cell Biol 166:369–380

Krueger A, Schmitz I, Baumann S, Krammer PH, Kirchhoff S (2001) Cellular FLICE-inhibitory protein splice variants inhibit different steps of caspase-8 activation at the CD95 death-inducing signaling complex. J Biol Chem 276:20633–20640

Lavrik I, Golks A, Krammer PH (2005a) Death receptor signaling. J Cell Sci 118:265–267

Lavrik IN, Golks A, Krammer PH (2005b) Caspases: pharmacological manipulation of cell death. J Clin Invest 115:2665–2672

Lavrik IN, Golks A, Riess D, Bentele M, Eils R, Krammer PH (2007) Analysis of CD95 threshold signaling: triggering of CD95 (FAS/APO-1) at low concentrations primarily results in survival signaling. J Biol Chem 282:13664–13671

Lee S-M, Kim E-J, Suk K, Lee W-H (2011) Stimulation of Fas (CD95) induces production of pro-inflammatory mediators through ERK/JNK-dependent activation of NF-κB in THP-1 cells. Cell Immunol 271(1):157–62

Legembre P, Barnhart BC, Peter ME (2004) The relevance of NF-κB for CD95 signaling in tumor cells. Cell Cycle 3:1235–1239

Legewie S, Blüthgen N, Herzel H (2006) Mathematical modeling identifies inhibitors of apoptosis as mediators of positive feedback and bistability. PLoS Comput Biol 2:e120

Li H (1998) Cleavage of BID by caspase 8 mediates the mitochondrial damage in the Fas pathway of apoptosis. Cell 94:491–501

Li Q, Verma IM (2002) NF-κB regulation in the immune system. Nat Rev Immunol 2:725–734

Macal CM, North MJ (2005) Tutorial on agent-based modeling and simulation. J Simul 4:151–162

Macal, C.M.; North, M.J.; , "Agent-based modeling and simulation", Proceedings of the 2009 Winter Simulation Conference (WSC), pp.86–98, 13–16Dec. 2009 doi:http://ieeexplore.ieee.org/stamp/stamp.jsp?tp=&arnumber=5429318&isnumber=5429163

Mai Z, Liu H (2009) Boolean network-based analysis of the apoptosis network: irreversible apoptosis and stable surviving. J Theor Biol 259:760–769

Medema JP, Toes RE, Scaffidi C, Zheng TS, Flavell RA, Melief CJ, Peter ME, Offringa R, Krammer PH (1997) Cleavage of FLICE (caspase-8) by granzyme B during cytotoxic T lymphocyte-induced apoptosis. Eur J Immunol 27:3492–3498

Micheau O, Thome M, Schneider P, Holler N, Tschopp J, Nicholson DW, Briand C, Grütter MG (2002) The long form of FLIP is an activator of caspase-8 at the Fas death-inducing signaling complex. J Biol Chem 277:45162–45171

Moquin D, Chan FK-M (2010) The molecular regulation of programmed necrotic cell injury. Trends Biochem Sci 35:434–441

Mühlethaler-Mottet A, Flahaut M, Bourloud KB, Nardou K, Coulon A, Liberman J, Thome M, Gross N (2011) Individual caspase-10 isoforms play distinct and opposing roles in the initiation of death receptor-mediated tumour cell apoptosis. Cell Death Dis 2:e125

Muzio M, Chinnaiyan AM, Kischkel FC, O'Rourke K, Shevchenko A, Ni J, Scaffidi C, Bretz JD, Zhang M, Gentz R et al (1996) FLICE, a novel FADD-homologous ICE/CED-3-like protease, is recruited to the CD95 (Fas/APO-1) death–inducing signaling complex. Cell 85:817–827

Nakajima A, Kojima Y, Nakayama M, Yagita H, Okumura K, Nakano H (2008) Downregulation of c-FLIP promotes caspase-dependent JNK activation and reactive oxygen species accumulation in tumor cells. Oncogene 27:76–84

Neumann L, Pforr C, Beaudouin J, Pappa A, Fricker N, Krammer PH, Lavrik IN, Eils R (2010) Dynamics within the CD95 death-inducing signaling complex decide life and death of cells. Mol Syst Biol 6:352

Oberst A, Pop C, Tremblay AG, Blais V, Denault J-B, Salvesen GS, Green DR (2010) Inducible dimerization and inducible cleavage reveal a requirement for both processes in caspase-8 activation. J Biol Chem 285:16632–16642

O'Reilly LA, Divisekera U, Newton K, Scalzo K, Kataoka T, Puthalakath H, Ito M, Huang DCS, Strasser A (2004) Modifications and intracellular trafficking of FADD/MORT1 and caspase-8 after stimulation of T lymphocytes. Cell Death Differ 11:724–736

Paulsen M, Janssen O (2011) Pro- and anti-apoptotic CD95 signaling in T cells. Cell Commun Signal 9:7

Paulsen M, Valentin S, Mathew B, Adam-Klages S, Bertsch U, Lavrik I, Krammer PH, Kabelitz D, Janssen O (2011) Modulation of CD4+ T-cell activation by CD95 co-stimulation. Cell Death Differ 18:619–631

Peter ME, Budd RC, Desbarats J, Hedrick SM, Hueber A-O, Newell MK, Owen LB, Pope RM, Tschopp J, Wajant H et al (2007) The CD95 receptor: apoptosis revisited. Cell 129:447–450

Rasper DM, Vaillancourt JP, Hadano S, Houtzager VM, Seiden I, Keen SL, Tawa P, Xanthoudakis S, Nasir J, Martindale D et al (1998) Cell death attenuation by 'Usurpin', a mammalian DED-caspase homologue that precludes caspase-8 recruitment and activation by the CD-95 (Fas, APO-1) receptor complex. Cell Death Differ 5:271–288

Rathmell JC, Cooke MP, Ho WY, Grein J, Townsend SE, Davis MM, Goodnow CC (1995) CD95 (Fas)-dependent elimination of self-reactive B cells upon interaction with CD4+ T cells. Nature 376:181–184

Rehm M, Huber HJ, Dussmann H, Prehn JHM (2006) Systems analysis of effector caspase activation and its control by X-linked inhibitor of apoptosis protein. EMBO J 25:4338–4349

Rehm M, Huber HJ, Hellwig CT, Anguissola S, Dussmann H, Prehn JHM (2009) Dynamics of outer mitochondrial membrane permeabilization during apoptosis. Cell Death Differ 16:613–623

Roche B, Drake JM, Rohani P (2011) An agent-based model to study the epidemiological and evolutionary dynamics of Influenza viruses. BMC Bioinformatics 12:87

Saez-Rodriguez J, Alexopoulos LG, Epperlein J, Samaga R, Lauffenburger DA, Klamt S, Sorger PK (2009) Discrete logic modeling as a means to link protein signaling networks with functional analysis of mammalian signal transduction. Mol Syst Biol 5:331

Scaffidi C, Medema JP, Krammer PH, Peter ME (1997) FLICE is predominantly expressed as two functionally active isoforms, caspase-8/a and caspase-8/b. J Biol Chem 272:26953–26958

Scaffidi C, Fulda S, Srinivasan A, Friesen C, Li F, Tomaselli KJ, Debatin KM, Krammer PH, Peter ME (1998) Two CD95 (APO-1/Fas) signaling pathways. EMBO J 17:1675–1687

Scaffidi C, Schmitz I, Krammer PH, Peter ME (1999) The role of c-FLIP in modulation of CD95-induced apoptosis. J Biol Chem 274:1541–1548

Schlatter R, Schmich K, Avalos Vizcarra I, Scheurich P, Sauter T, Borner C, Ederer M, Merfort I, Sawodny O (2009) ON/OFF and beyond–a boolean model of apoptosis. PLoS Comput Biol 5: e1000595

Schneider P (1998) Conversion of membrane-bound Fas(CD95) ligand to its soluble form is associated with downregulation of its proapoptotic activity and loss of liver toxicity. J Exp Med 187:1205–1213

Schug ZT, Gonzalvez F, Houtkooper RH, Vaz FM, Gottlieb E (2011) BID is cleaved by caspase-8 within a native complex on the mitochondrial membrane. Cell Death Differ 18:538–548

Schrodinger, LLC (2010). The PyMOL Molecular Graphics System, Version 1.3r1

Scott FL, Stec B, Pop C, Dobaczewska MK, Lee JJ, Monosov E, Robinson H, Salvesen GS, Schwarzenbacher R, Riedl SJ (2009) The Fas-FADD death domain complex structure unravels signaling by receptor clustering. Nature 457:1019–1022

Sen R, Baltimore D (1986) Inducibility of κ immunoglobulin enhancer-binding protein NF-κB by a posttranslational mechanism. Cell 47:921–928

Shi Y (2002) Mechanisms of caspase activation and inhibition during apoptosis. Mol Cell 9:459–470

Shikama Y, Yamada M, Miyashita T (2003) Caspase-8 and caspase-10 activate NF-κB through RIP, NIK and IKKalpha kinases. Eur J Immunol 33:1998–2006

Spencer SL, Sorger PK (2011) Measuring and modeling apoptosis in single cells. Cell 144:926–939

Sprick MR, Rieser E, Stahl H, Grosse-Wilde A, Weigand MA, Walczak H (2002) Caspase-10 is recruited to and activated at the native TRAIL and CD95 death-inducing signaling complexes in a FADD-dependent manner but cannot functionally substitute caspase-8. EMBO J 21:4520–4530

Steller EJ, Ritsma L, Raats DA, Hoogwater FJ, Emmink BL, Govaert KM, Laoukili J, Rinkes IH, van Rheenen J, Kranenburg O (2011) The death receptor CD95 activates the cofilin pathway to stimulate tumour cell invasion. EMBO Rep 12(9):931–937

Strasser A, Jost PJ, Nagata S (2009) The many roles of FAS receptor signaling in the immune system. Immunity 30:180–192

Tang D, Lotze MT, Kang R, Zeh HJ (2010) Apoptosis promotes early tumorigenesis. Oncogene 30(16):1851–4

Tenev T, Bianchi K, Darding M, Broemer M, Langlais C, Wallberg F, Zachariou A, Lopez J, Macfarlane M, Cain K, Meier P (2011) The ripoptosome, a signaling platform that assembles in response to genotoxic stress and loss of IAPs. Mol cell 43(3):432–48

Trauth BC, Klas C, Peters AM, Matzku S, Möller P, Falk W, Debatin KM, Krammer PH (1989) Monoclonal antibody-mediated tumor regression by induction of apoptosis. Science 245:301–305

Ueffing N, Schuster M, Keil E, Schulze-Osthoff K, Schmitz I (2008) Up-regulation of c-FLIP short by NFAT contributes to apoptosis resistance of short-term activated T cells. Blood 112:690–698

van Raam BJ, Salvesen GS (2011) Proliferative versus apoptotic functions of caspase-8 hetero or homo: the caspase-8 dimer controls cell fate. Biochim Biophys Acta 1824(1):113–22

Vandenabeele P, Declercq W, Van Herreweghe F, Vanden Berghe T (2010a) The role of the kinases RIP1 and RIP3 in TNF-induced necrosis. Sci Signal 3(115):re4

Vandenabeele P, Galluzzi L, Vanden Berghe T, Kroemer G (2010b) Molecular mechanisms of necroptosis: an ordered cellular explosion. Nat Rev Mol Cell Biol 11:700–715

Verma IM, Stevenson JK, Schwarz EM, Van Antwerp D, Miyamoto S (1995) Rel/NF-κ B/I kappa B family: intimate tales of association and dissociation. Genes Dev 9:2723–2735

Wang L, Yang JK, Kabaleeswaran V, Rice AJ, Cruz AC, Park AY, Yin Q, Damko E, Jang SB, Raunser S et al (2010) The Fas-FADD death domain complex structure reveals the basis of DISC assembly and disease mutations. Nat Struct Mol Biol 17:1324–1329

Wong WW-L, Gentle IE, Nachbur U, Anderton H, Vaux DL, Silke J (2010) RIPK1 is not essential for TNFR1-induced activation of NF-κB. Cell Death Differ 17:482–487

Yan N, Shi Y (2005) Mechanisms of apoptosis through structural biology. Annu Rev Cell Dev Biol 21:35–56

Yang ZR (2005) Prediction of caspase cleavage sites using Bayesian bio-basis function neural networks. Bioinformatics 21:1831–1837

Yu J, Zhang L (2005) The transcriptional targets of p53 in apoptosis control. Biochem Biophys Res Commun 331:851–858

Yu JW, Jeffrey PD, Shi Y (2009) Mechanism of procaspase-8 activation by c-FLIPL. Proc Natl Acad Sci USA 106:8169–8174

Chapter 3
Systematic Complexity Reduction of Signaling Models and Application to a CD95 Signaling Model for Apoptosis

Dennis Rickert, Nicolai Fricker, Inna N. Lavrik, and Fabian J. Theis

Abstract A major problem when designing mathematical models of biochemical processes to analyze and explain experimental data is choosing the correct degree of model complexity. A common approach to solve this problem is top-down: Initially, complete models including all possible reactions are generated; they are then iteratively reduced to a more manageable size. The reactions to be simplified at each step are often chosen manually since exploration of the full search space seems unfeasible. While such a strategy is sufficient to identify a single, clearly structured reduction of the model, it discards additional information such as whether some model features are essential. In this chapter, we introduce alternate set-based strategies to model reduction that can be employed to exhaustively analyze the complete reduction space of a biochemical model instead of only identifying a single valid reduction.

D. Rickert • F.J. Theis (✉)
Institute of Bioinformatics and Systems Biology, Helmholtz Zentrum München, Ingolstädter Landstraße 1, 85764 Neuherberg, Germany
e-mail: fabian.theis@helmholtz-muenchen.de

N. Fricker
Institute of Bioinformatics and Systems Biology, Helmholtz Zentrum München, Ingolstädter Landstraße 1, 85764 Neuherberg, Germany

Division of Immunogenetics, German Cancer Research Center (DKFZ), Heidelberg 69120, Germany

Bioquant, Heidelberg 69120, Germany

I.N. Lavrik
Department of Translational Inflammation Research, Institute of Experimental Internal Medicine, Otto von Guericke University, Leipziger Str. 44, 39120 Magdeburg, Germany
e-mail: inna.lavrik@med.ovgu.de

I.N. Lavrik (ed.), *Systems Biology of Apoptosis*, DOI 10.1007/978-1-4614-4009-3_3,

3.1 Introduction

A major problem when designing mathematical models of biochemical processes to analyze and explain experimental data is choosing the correct degree of model complexity. Minimalistic models that include only the core reactions of a regulatory pathway will often fail to capture all mechanisms and be unable to reproduce the experimentally observed dynamics. In contrast, models that include all possible interactions may suffer from overfitting, strongly diminishing their predictive and analytical value. This becomes especially severe in pathways where interacting molecules are modified by, or bind with multiple interaction partners, as is common in inter- and intracellular signaling.

A common approach to solve this problem is top-down: Initially, complete models including all possible reactions are generated; they are then iteratively reduced to a more manageable size. The reactions to be simplified at each step are often chosen manually since exploration of the full search space seems unfeasible. While such a strategy is often sufficient if the goal is limited to finding a single, clearly structured reduction of the model, other questions that could be of interest from a modeling point of view are not considered. Examples for such questions are:

- Are some model features essential, i.e., can never be reduced?
- What are the smallest versions of the model that are still viable?
- Is there a logical pattern common to all valid model reductions?

These questions will often require the analysis of all possible model reductions, a task that is both time consuming and repetitive, making it ill-suited to manual analysis. Explicit enumeration and testing of all possible reductions is prohibitively expensive, imposing the need to utilize heuristic search strategies. In this chapter we will discuss a set-based strategy that is suited to answer the questions posed above. While we explain the strategy in the context of ODE modeling with mass action kinetics, it can be adapted to a wide range of models, including SDEs, Boolean models, and stochastic simulations. We finish the chapter with an application to CD95 signaling.

3.2 Graphical Structuring of ODE Models

It is often helpful to organize ODE models into a graph-like data structure. Not only does this allow easy visualization of the model, it also allows the utilization of established graph-based operations, like the removal of edges, and graph-based algorithms, e.g., connectivity analysis with little adaption. This is frequently done by researches, however, depending on the exact application, the details of the representation differ. To avoid confusion, we give a short overview of the approach we utilize. An example for a mass action ODE that is represented by a graph is illustrated in Fig. 3.1. Note that while we limit ourselves to mass action kinetics in this chapter to avoid excessively complex notation, the general approach is also valid for more complex reaction kinetics, such as Michaelis–Menten kinetics.

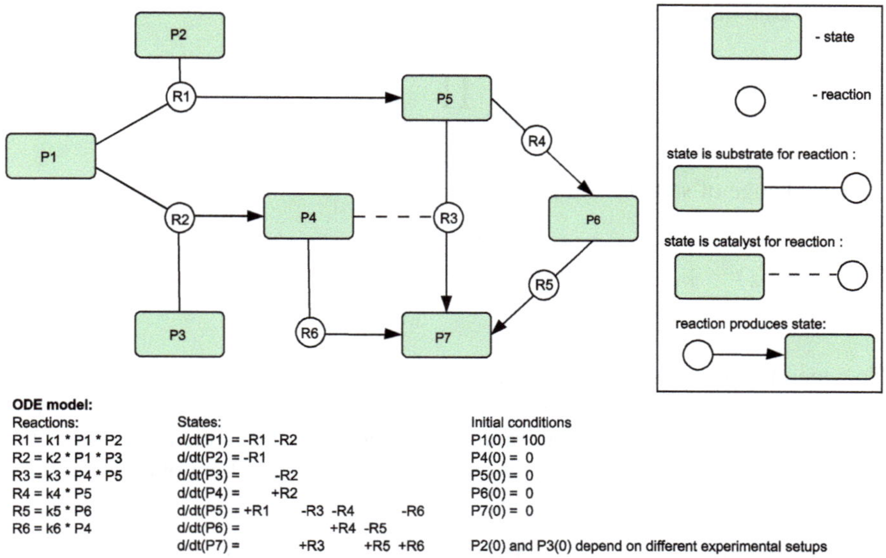

Fig. 3.1 A graphically represented ODE model. An ODE model represented as a bipartite graph. *Green boxes* are states (usually biochemical molecules), *white circles* are reactions. A state can be either a substrate (*normal line*) or a catalyst (*dashed line*) for a reaction. A directed line from a reaction to a state indicates that the reaction produces this state. In this model, the state P1–P3 act as input (e.g., signaling molecules, primary metabolites...), the state P7 is the final model output. P1 dimerizes with either P2 or P3. Both the P1–P2 dimer (P5) and the P1–P3 dimer (P4) can be activated independently into P7 (P5 via the intermediate P6, P4 directly). In addition, there is a cooperative mode of activation where P4 catalyzes the activation of P5

We focus on ODE models that can be decomposed into two types of elements: *states* and *reactions*. States represent the molecular species of the system we want to analyze, whereas reactions are the biochemical reactions that change one or more molecule into other products. For a given ODE model, we will usually know the value of all reactions depending on the current state values and the first derivate of all states depending on the current reaction values. This is sufficient to simulate the model using a numerical solver.

Based on the following definitions:

$S_{1...k}$	The set of all states
$R_{1...j}$	The set of all reactions
k_n	The kinetic rate of reaction n.
$\text{Subs}_R(R_n)$	is the index set of all states that are substrate of reaction n.
$\text{Subs}_S(S_n)$	is the index set of all reactions that consume state n.
$\text{Cat}_R(R_n)$	is the index set of all states that are catalysts of reaction n.
$\text{Cat}_S(S_n)$	is the index set of all reactions for which state n is a catalyst.
$\text{Prod}(S_n)$	is the index set of all reactions that produce state n.
$\text{stoich}(S_i, R_n)$	is the stoichiometric constant of state i in reaction n (the stoichiometric constant denotes how many molecules of each type participate in a reaction, e.g., for a homodimerization the stoichiometric constant is 2).

The value of reaction n at time t:

$$R_n = k_n \prod_{i \in (\mathrm{Subs}_R(R_n) \cup \mathrm{Cat}(R_n))} S_i^{\mathrm{stoich}(S_i, R_n)}$$

The change of state m at time t:

$$\frac{\mathrm{d}}{\mathrm{d}t}(S_n) = + \sum_{i \in \mathrm{Prod}(S_n)} \mathrm{stoich}(S_n, R_i) * R_i - \sum_{i \in \mathrm{Subs}_S(S_n)} \mathrm{stoich}(S_n, R_i) * R_i$$

To generate the model graph for a given ODE:

For a graphical representation of this an ODE system, both states and reactions are considered nodes in a model graph. The model graph is bipartite graph, i.e., there will only be connections between reactions and states, but not between states and states or reactions and reactions.

- There is a directed edge from every state n to every reaction m if $n \in \mathrm{Subs}_R(R_m)$, i.e., if state n is substrate for reaction. We label all these edges with stoich (S_n, R_m) and color them as substrate influences.
- There is a directed edge from every state n to every reaction m if $n \in \mathrm{Cat}(R_m)$, i.e., if state n catalyzes reaction m. We label all these edges with $\mathrm{stoich}(S_n, R_m)$ and color them as catalytic influences.
- There is a directed edge from each reaction m to each state n if $m \in \mathrm{Prod}(S_n)$, i.e., if reaction m produces state n. We label all these edges with $\mathrm{stoich}(S_n, R_m)$ and color them as productions.

It is easy to see that additional, more complex kinetics can be supported by simply increasing the number of different colors used to color the edges from states to reaction nodes.

3.2.1 Experimental Noise Model and Error Function

One of the most important tasks in model-driven systems biology is to evaluate how well a model explains experimental observation. Typically, we will want to run simulations of different experimental conditions, such as different intensities of stimulation and compare the resulting model dynamics to experimental measurements. Figure 3.2 illustrates how the simulation of biochemical models produces time course data for different experimental conditions.

While generating the experimental data is usually straightforward and can be done using established toolboxes such as the SBtoolbox 2 for Matlab (Schmidt 2006), the comparison is less straightforward and can be influenced by personal bias. To minimize this subjectivity, it is useful to quantify the difference between model and experimental data. This is often done by utilizing *error functions*. An error function quantifies the difference between time course data

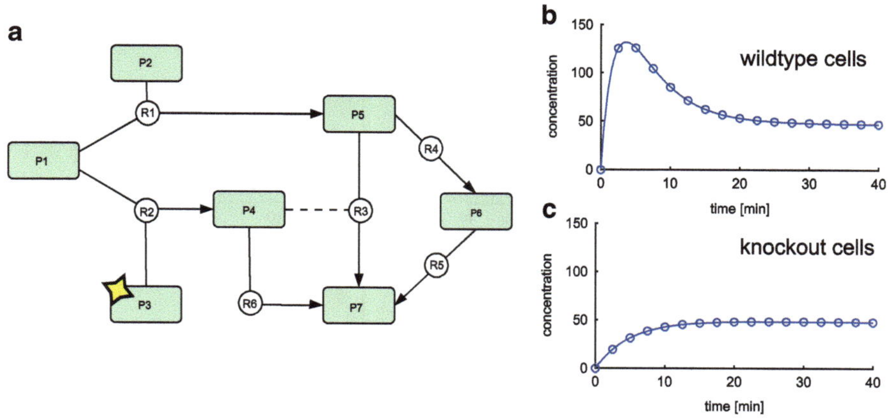

Fig. 3.2 The behavior of dynamic systems can be changed by different experimental conditions. (**a**) A typical experimental setup will often involve the reduction or complete knockout of one or more states of a dynamical system. In our example, the protein P3 is completely knocked out (set to concentration zero). (**b**) In wild-type cells, the reaction to stimulation is an early peak in activity (at around 5 min), followed by a slow decay to a constant concentration level of about 50. (**c**) In knockout cells the early peak at 5 min is missing. The later activation remains unaffected by the knockout

generated by the model and experimental observations. However both the generation and the interpretation of error functions and their values are nontrivial.

A major problem in systems biology is that data will often be very noisy; even repeated measurements of the same system at the same time can vary by 10% or more. This is caused both by measurement errors and by the high variability of biological systems. In contrast deterministic models such as ODEs will always reproduce identical results without any noise; therefore, it is unrealistic to expect an ODE to reproduce experimental observations perfectly.

To deal with this situation, we try to capture the experimental variance in an error noise model. For many different experimental setups, we observe that measurement errors and experimental variance are normally distributed $N(\mu, \sigma^2)$ with a mean μ of zero and a variance σ^2 that depends on the exact experimental setup. We assume that the "basic" behavior of the system is determined by the deterministic model and that the difference between deterministic model and observed data is caused by the experimental variance. This is illustrated in Fig. 3.3.

$$\begin{aligned}\text{Observation} = {} & \text{deterministic behavior (ODE model)} \\ & + \text{model variance (stochastic component)}\end{aligned}$$

Based on this assumption we can calculate how probable it is to observe our experimental data. Initially we will only consider the case where a single variable (e.g., protein concentration) is observed at the time points $1 \ldots n$. This is mainly done to avoid confusing notation, in the end we will derive the total error value by simply summing over all individual error values. Let $x^{\text{obs}} = x^{\text{obs}}_{1\ldots n}$ be the vector of our experimental observations if only a single replicate of the experiment is performed.

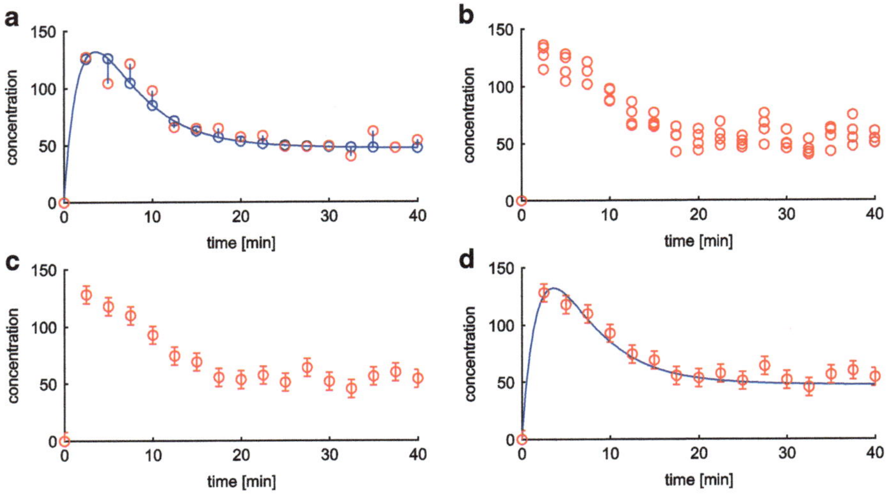

Fig. 3.3 Experimental observations of dynamic systems can often be decomposed in deterministic and stochastic components. (**a**) Time course data from a completely deterministic system, e.g., either a biological system with little variance or an artificial model (*blue line*) and the corresponding measurements with added stochastic effects (*red circles*). (**b**) To deal with stochastic effects, multiple measurements of each time point are performed, resulting in multiple values for each time point (*red circles*). (**c**) To quantify multiple replicates of experimental measurements, the mean value (*circle*) and standard deviation (*error bars*) for each time point are determined. (**d**) Comparison of the deterministic time course (*blue*) and the quantification of multiple observations (*red*)

If the experiment is repeated multiple times, x^{obs} instead contains the mean value of all observations. The deterministic time course data produced by simulating our ODE model is $x^{\text{sim}} = x^{\text{sim}}_{1\ldots n}$ and $x^{\text{stoch}} = x^{\text{stoch}}_{1\ldots n}$ is the difference between observation and simulation that we attribute to stochastic effects in our experimental setup:

$$\begin{aligned} x^{\text{obs}}_{1\ldots n} &= x^{\text{sim}}_{1\ldots n} + x^{\text{stoch}}_{i} \\ x^{\text{stoch}}_{i} &= x^{\text{obs}}_{1\ldots n} - x^{\text{sim}}_{1\ldots n} \end{aligned}$$

with $x^{\text{stoch}}_{i} \sim N(\mu, \sigma_i^2)$ and $\mu = 0$ we find (without proof) that the probability for every single observation is

$$P\left(x^{\text{stoch}}_{i}\right) = \frac{1}{\sqrt{2\pi\sigma_i^2}} \mathrm{e}^{-\frac{\left(x^{\text{stoch}}_{i}\right)^2}{2\sigma_i^2}}$$

If we assume independence between the experimental variations at different time points (e.g., that a large variation at an early time point is not the cause of large variations at later time points), we can derive the probability for the entire time course by multiplying the probabilities of each single time point:

$$P\left(x^{\text{stoch}}\right) = \prod_{i=1}^{n} P\left(x^{\text{stoch}}_{i}\right) = \prod_{i=1}^{n} \frac{1}{\sqrt{2\pi\sigma_i^2}} \mathrm{e}^{-\frac{\left(x^{\text{stoch}}_{i}\right)^2}{2\sigma_i^2}}$$

$P(x^{\text{stoch}})$ is already a quantification of how well our model fits the experimental data; however, in practice this formula is rather inconvenient as it requires multiple evaluations of the exponential function and generally results in extremely small values that are too small to be represented in standard computational variables. However we can simplify the equation by dropping the normalizing factor $1/\sqrt{2\pi\sigma_i^2}$ and rescaling it on a logarithmic scale. Logarithmic rescaling allows us to remove the exponential function and replace the product over each time point with a sum. In addition, as we are generally interested in an equation that directly evaluates the quality of the simulation $x_{1\dots n}^{\text{sim}}$, we substitute $x_i^{\text{stoch}} = x_{1\dots n}^{\text{obs}} - x_{1\dots n}^{\text{sim}}$ and derive the error function $\text{EF}(x^{\text{sim}})$:

$$\text{EF}(x^{\text{sim}}) = \sum_{i=1}^{n} \left[\frac{\left(x_i^{\text{obs}} - x^{\text{sim}}\right)^2}{2\sigma_i^2} \right]$$

This formula is the sum of squared residuals, normalized by the variance of the observation. The value of the error function is zero if $x^{\text{obs}} = x^{\text{sim}}$ and positive otherwise; the closer the value of the error function is to zero, the better our simulated data fits to our experimental observations. If multiple replicates of the experiment have been performed, the experimental variance σ_i^2 can be calculated directly from the data points; otherwise it has to be estimated based on expert knowledge.

3.2.2 Parameter Optimization

Based on our introduction to ODE's in Sect. 3.2, it is obvious that the value of the simulated time course x^{sim} depends on a set of parameters. We will group all parameters of the ODE system into a vector of parameters we will call θ, where θ_i is the i-th parameter. For a mass action kinetic model, θ will usually consist of the kinetic rates of all reactions.

We will utilize the error function introduced in the previous section to find a value for θ that results in a good fit between simulation and experimental data. We do this by trying to minimize the value of $\text{EF}(x^{\text{sim}}(\theta))$. This process is called *parameter optimization*. As entire books have been written on the topic, we will limit us here to an overview over a few common techniques. All approaches introduced here are based on the same general idea; an initial value for θ is picked (either based on literature values or at random) and subsequently modified with the goal of improving the error function value. This is done iteratively until some kind of ending criteria is met. Common ending criteria include a fixed number of total iterations or a number of iterations without marked improvement in function value.

A very basic strategy is the *hill climbing* algorithm. In each step the local neighborhood of the current θ is explored. To do so, a candidate for a new parameter set, $\hat{\theta}$ is generated by adding a value that depends on the exact implementation of the algorithm to a single element θ_i. If the change results in an improvement, θ is updated to the new value $\hat{\theta}$. Otherwise another candidate is generated by subtracting the same value from θ_i and again, the new candidate is accepted if it results in an improvement over the previous error value. This is then iteratively repeated for each element of θ until no further improvement is made. While this method is easy to implement, it has the disadvantage of frequently becoming stuck in local optima.

Another type of local methods are *steepest ascent methods*. In these methods, the next step in each iteration is chosen based on local evaluation or approximation of the first derivative of the model dynamics. This will usually result in a step that optimizes the improvement of the error function. While this sounds like a promising approach, it is limited by how well we can approximate the first derivate of the model. In addition, models that are determined by higher-order derivates will result in very small steps, causing long run times. As in hill climbing algorithms, there is a danger of getting stuck in local minima.

A common heuristic to deal with the issue of local optima are *simulated annealing* algorithms. The idea of these algorithms is that in each step a proposal is generated by randomly changing the current parameter vector. If the change results in a score improvement, it is always accepted. However, if the error value increases, the proposal is still accepted with a certain probability. This probability depends on a temperature value that start at high value and is then decreased according to a cooling schedule.

3.2.3 Choosing Significant Error Function Cutoffs

In the previous sections we discussed how to quantify the difference between the simulated time course data and the experimental observations, and gave an overview over parameter optimization techniques that can be used to minimize this quantified value in order to produce a good model fit. However, when we try to reduce models, we face the question whether a slightly worse error score justifies a significant simplification of the model. This leads to the question of *cutoff values* of the error function: Up to which error value can we say that a model reproduces our experimental data satisfactory?

One way to derive cutoffs for error values is based on the assumption that the observed data points are normally distributed around a time course generated by a deterministic dynamic. This is done using the χ^2 *distribution*. The χ^2 distribution calculates the probability that summing over a number k of squared, uniform normally distributed random variables results in a certain value. If we keep in mind that we assumed $(x_i^{\text{obs}} - x^{\text{sim}})$ to be normally distributed, this is exactly what we do in the error function $\text{EF}(x^{\text{sim}}) = \sum_{i=1}^{n} \left[\frac{\left(x_i^{\text{obs}} - x^{\text{sim}}\right)^2}{2\sigma_i^2} \right]$.

Based on the χ^2 distribution, we can estimate the *expected deviation* of an observation from the real value of the generating system and the resulting *expected error score per time point*. For example, we expect that 68% of all data points should be within one standard deviation of the deterministic time course (contributing an error value less than one error unit per time point) and 95% should be within two standard deviations (contributing an error value less than four error units per time point).

The exact calculation and interpretation of confidence intervals using this method is nontrivial and exceeds the scope of this chapter. However, as *a rule of thumb* based on these considerations, we expect a normalized sum of squared residuals lesser or *equal to the number of time points* to be almost always a rather good fit that explains most data points. Likewise, a score *larger than four times the number of time points* is almost always a bad fit that either completely misses some data points or shows a significant deviation from every single measurement.

It should be noted that, no matter how error cutoff values are derived, they should always be analyzed in the context of the experimental data and the biological system. It is very possible that a way to derive a cutoff value that works perfectly well for one set of data results in a cutoff that is to strict or to permissive in a different context. It generally makes sense to test multiple cutoff values and compare which value is closest to the interpretation of the experimental data in a biological context (Fig. 3.4).

3.3 Reduction of Graph-Based Models

In the context of this chapter, we focus on reducing a model by removing reactions that are not required to explain the observed dynamic behavior of the model. This reduces the degree of parameter under determination and helps to identify the core dynamics essential to the model.

In general, two different types of reactions compose a biochemical model. One is a set of *core reactions* that can be considered essential for a pathway. These reactions have either been confirmed in previous experiments, are established as a gold standard in literature, or are required in the model for structural reason, e.g., reactions that are important for model connectivity. We are generally not interested in reducing a model by removing core reactions.

In addition there are *auxiliary reactions* or *reduction candidates* R_{red}, reactions that are either of a hypothetical nature or of a detail level that might be inappropriate for the desired model. Examples for reduction candidates often include reactions that have been predicted based on binding domain analysis and PPI data, but have not been confirmed in vivo. For our analysis, any subset of R_{red} is a potential reduction of the model. The set of all subsets of a set X is also called the powerset of X, which we will denote $\mathrm{P}(X)$. Thus the entire space of possible reductions S_{red} is $P(R_{\mathrm{red}})$. The size of a powerset grows exponentially to the

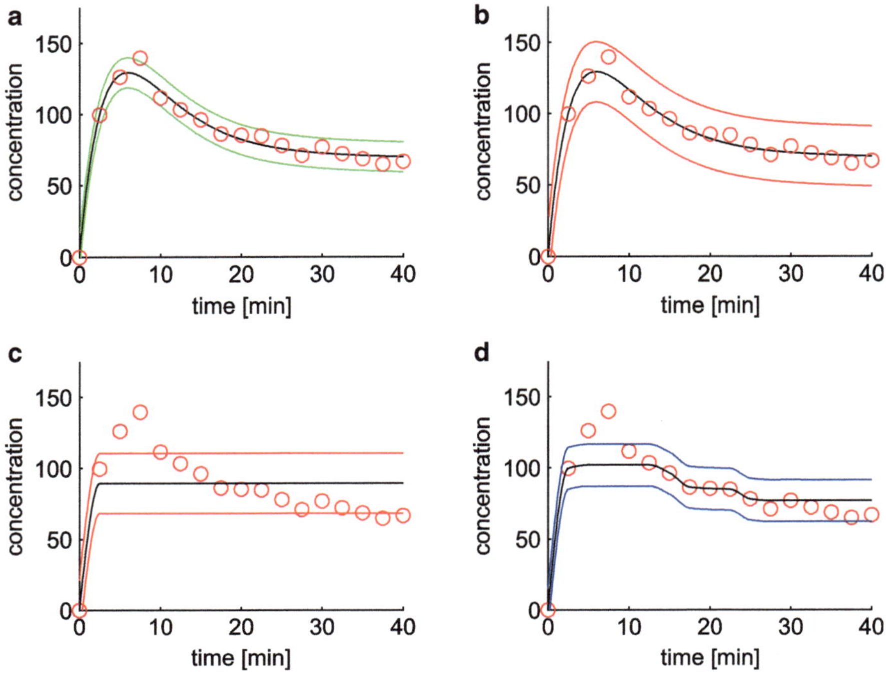

Fig. 3.4 Visualization of different kinds of cutoffs. (**a**) Cutoff based on one standard deviation (cutoff value equal to the number of time points). The *black line* is a possible simulation that satisfies the cutoff value, the green lines visualize the simulation +/− one standard deviation. Note that the borders visualized with respect to the simulated data; this is functionally identical to considering the borders with respect to the experimental data points. Note that it is not necessary that all points have to be inside visualized borders; if some data points are close to the *black line*, others might be outside the one standard deviation border. (**b**) Cutoff based on two standard deviations (cutoff value equal to four times the number of time points). The *red lines* visualize a two standard deviation border, (**c**) Cutoff based on two standard deviations, worst case fit. The simulated time course still results in an error value below the cutoff, despite two data points being outside *red borders*. In this case, the two standard deviation cutoff is too permissive, as an important qualitative feature of the data, the early activity peak is lost. (**d**) By analyzing the minimal error score at which the peaking behavior is lost, we derive a new error cutoff based on 1.4 standard deviations. The new worst case fit (*black line*) still shows a clear early peak

base 2. This means that the total number of possible reductions of a model is 2^n, where n is the number of reactions in R_{red}.

Most times we find that not all members of S_{red} are valid reductions, i.e., able to reproduce experimental data with a quality below a cutoff as explained in Sect. 3.2.3. The challenge lies in identifying which reduction candidates are not *supported by experimental evidence*.

It should be noted that parameter optimization for an ODE model is often a computationally expensive task that can take multiple minutes per attempt. It is therefore necessary to keep direct testing of reductions to a minimum. Instead, we hope to verify/reject a large number of reductions *indirectly*. Since the complete reduction space grows exponentially with the number of auxiliary reductions, brute force checking of every reduction candidate will often take a prohibitively long time. *Heuristic strategies* need to be employed to speed up the identification reductions that are not supported by experimental evidence.

3.3.1 Indirect Model Verification and Rejection

Indirect model acceptance and rejection are based on a simple but powerful property resulting from the definition of our reduction framework. Based on the definitions given in Sect. 3.2, we find that removing a reaction from a model is identical to setting the associated kinetic parameter to zero. This implies that removing a reaction from an invalid model cannot transform the model into a valid model. Likewise adding reactions to a valid model cannot make this model invalid, as the newly added reactions could potentially have a kinetic rate of zero. This results in the following theorem:

Core theorem of set-based model reduction:

1. If a reduction is identified as valid, all reductions that are subsets of the valid reduction are also valid.
2. If a reduction is identified as invalid, all reductions that are supersets of the invalid reduction are also invalid.

These properties are essential to our design of reduction strategies. They allow us to accept valid and reject invalid reductions without the need to explicitly testing them. This is necessary, as the complete search space doubles with each additional reduction candidate, making explicit testing of all reductions impossible. We need to maximize the information gained indirectly in order to deal with the exponentially growing search space.

If we compare the indirect information gained from accepting/rejecting a reduction, we find that these will strongly differ between different candidates. Most of the time, reduction with few elements provide the most indirect information gain if they are rejected, as they are subsets of a larger number of reductions than large reductions. In contrast, large reductions provide the most information if they are accepted.

Direct testing of a reduction candidate is based on the analysis introduced in Sect. 3.2.2, 3.2.3, e.g., multiple parameter fitting attempts are started that attempt to find a model parameterization that explains the observed data with a quality below a cutoff. While we only introduce cutoffs that are motivated by statistical analysis of a given error model, cutoffs and error functions derived in a different way can also be utilized without changing the other aspects of model reduction. If a parameterization is found that results in an error value below a certain cutoff, the reduction is accepted as valid.

Unfortunately, we often find that the probability of accepting a large reduction picked at random is rather small; likewise, small reductions are more likely to be accepted than large reductions. This implies that, in order to analyze the reduction space efficiently, we need to focus on strategies that identify large reductions that are likely to be accepted or small reductions that are likely to be rejected with a higher than random frequency.

In addition to estimating the probability of rejecting/accepting a reduction, a second important aspect to optimize indirect information gain is keeping track of the reductions that have already been accepted and rejected. While accepting a large reduction will often result in a significant information gain, this is only true if only a small number of its subsets have already been accepted. In contrast, if most subsets of a large reduction candidate have already been accepted, the information gained by accepting the candidate is still small.

3.4 Topological Model Analysis

As already mentioned, the reduction space we wish to analyze contains all combinatorial subsets of the reduction candidates of a model. This space will frequently contain reductions that can be identified as unable to reproduce experimental data based solely on the topology of the reduced model (e.g., cases where biologically important states become disconnected from the rest of the model). Examples for such reductions are illustrated in Fig. 3.5. We call these reductions *topologically invalid* reductions.

In addition, it is possible that different sets of elementary reductions result in models that show identical dynamic behavior. For these *redundant* models, it is sufficient to test the validity of one reduction of the redundancy group and subsequently assign all other models the same validity. Examples for redundant models are given in Fig. 3.5.

To recognize topologically invalid and redundant models we use the concepts of *observability* and *controllability* and *activity*. While these properties are inspired by the concepts with the same names as utilized in systems engineering, it should be noted that we use significantly different versions. We illustrate these concepts for the ODE models, but they can be applied in a similar way to a large range of different models, including SDE models, Boolean models, and agent-based models.

3.4.1 Controllability

We use the property *controllability* to keep track of which inputs are able to control which intermediate- and output states. An input is said to control another state if changing the input results in a change of behavior for the second state. An example

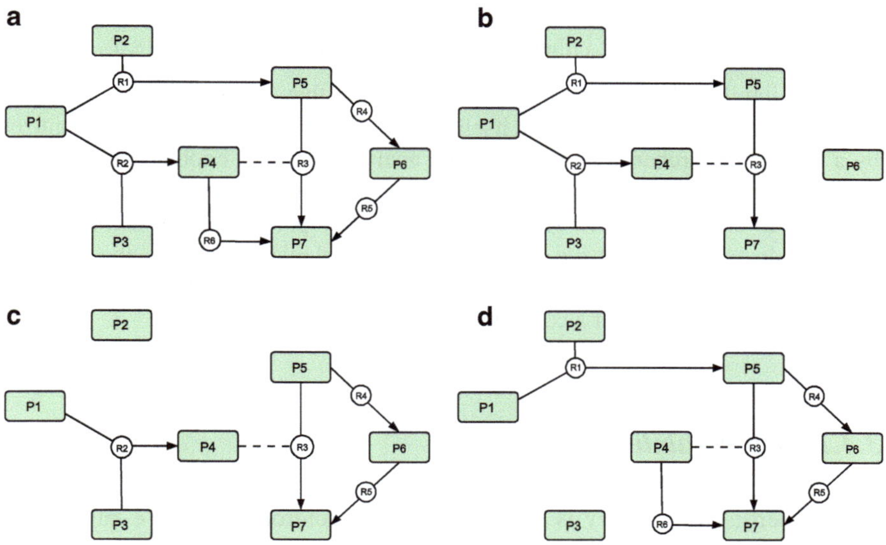

Fig. 3.5 Examples for valid, invalid, and redundant reductions. (**a**) is the original, unreduced model as introduced in Fig 3.1. (**b**) is a topologically valid reduction. Although both independent pathways that lead to the production of P7 have been removed, P7 is still produced by the cooperative pathway (reaction R3). If the cooperative reaction happens at a significantly higher rate than both indirect pathways, such a reduction can be realistic in a biological context. If this is the case, the model behavior is determined primarily by the cooperative activation mechanism. (**c**) is an invalid reaction. The direct activation of P4 has been removed. In addition, the dimerization of P1–P2 has been removed, so that P5 is no longer produced. This results in a situation where P7 can no longer be produced, rendering the entire pathway inactive. Therefore, model (**c**) is an invalid reduction. (**d**) This reduction is still able to produce P7 through the independent activation of P5, but can no longer produce state P4. Thus it is a valid reduction. However, it still includes two reactions that depend on P4 (R3 and R5). These reactions will never have a value larger zero, rendering them obsolete. A reduction of model (**d**) that would also remove R3 and R5 would act identically to model (**d**) without obsolete reactions. Therefore, we call model (**d**) redundant

for this is that during a gene knockout experiment, all other genes that are up- or down regulated are controlled by the gene knocked out. Models that do not connect the genes we observe to be experimentally controlled by the gene knocked out can be rejected without time-consuming parameter optimization attempts. For the type of ODE models that we consider, the control of states is *propagated* between the states by the reactions, as states do not directly depend upon each other. Reactions in turn are completely determined by a set of states that is specific for each reaction.

Intuitively, we find that an Input State SI:

- Controls itself by definition.
- Controls a reaction R_n if $(\mathrm{Subs}_R(R_n) \cup \mathrm{Cat}_R(R_n))$ contains any controlled states, i.e., if any substrate or catalyst of is R_n controlled.

- Controls a state S_n if $(\text{Subs}_S(S) \cup \text{Prod}(S_n))$ contains any controlled reactions, i.e., if any reaction that either consumes or produces S_n is controlled.

These intuitive definitions are lacking in so far as they allow for a state's controllability to depend recursively upon itself. While such a situation is easily recognized by a human researcher, a computer-based analysis needs to explicitly account for this possibility. It should be noted that these conditions are only *necessary*, but not *sufficient* to confirm control. As we only consider topological criteria, parameterizations can exist for which we do not observe control relations despite the topological conditions being fulfilled. This is the case if, e.g., the kinetic parameter of a reaction that is required for a control relation is set to zero.

3.4.2 Observability

The property *observability* is closely related to controllability. We call a state observable if changes to the state (either at a certain time point or to the initial conditions) can be recognized in the states that we are able to measure experimentally. Similar to the limitations imposed in the analysis of controllability, we are again limited to necessary conditions and have to avoid recursive dependency. We will use observability to identify reactions that are unimportant for the model dynamics we observe and can be removed. If any reactions are identified that are not observable, the model is by definition *redundant*, as a model that removes these would behave identically with respect to our experimental observations.

A state S_n:

- Is observable by definition if S_n is an output state
- Is observable if $(\text{Subs}_S(S_n) \cup \text{Cat}_S(S_n))$ contains any observable reaction

A reaction R_n:

- Is observable if $(\text{Subs}_R(R_n) \cup \text{Prod}(R_n))$ contains any observable state

3.4.3 Activity

We find that simulations of biological processes frequently contain only a few states that start with an initial concentration greater zero, whereas the majority of states will have an initial condition of zero. This can result in situations where multiple states and reactions will always have a concentration of zero, for all possible experimental setup. We use the property of *activity* to determine if a reduced model contains any reactions that will always have a value of zero. If this is the case, the model is *redundant*. In addition, if any output state is not active, the model is *invalid* (Figs. 3.6 and 3.7).

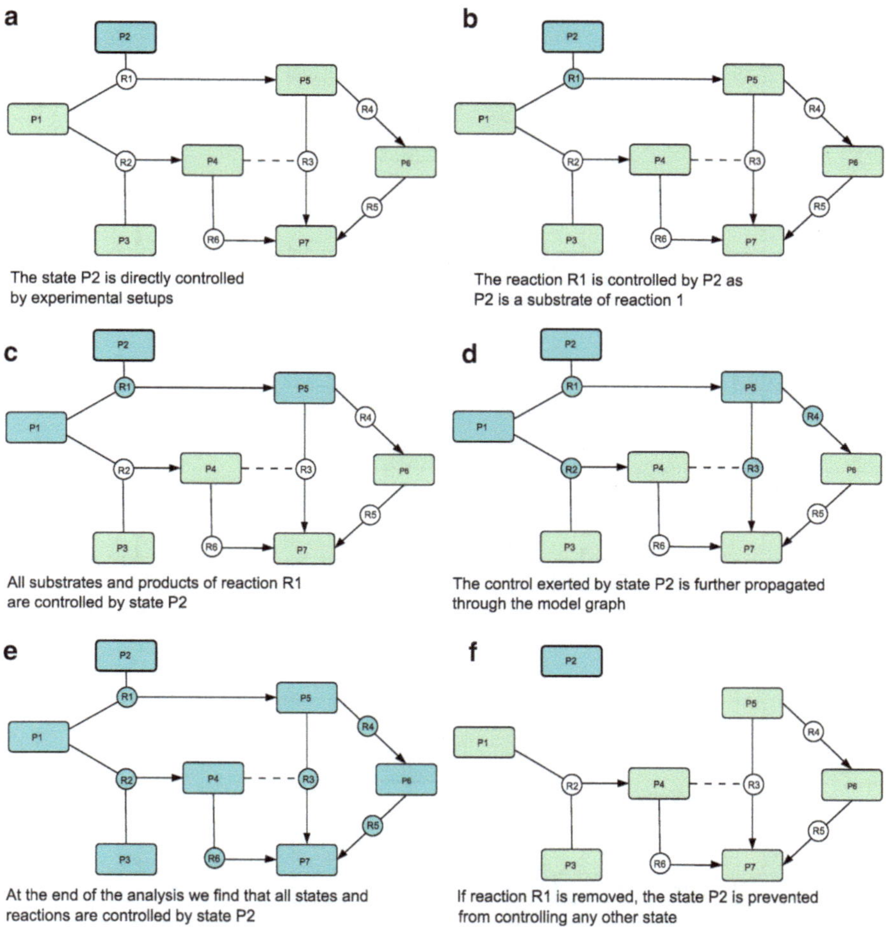

Fig. 3.6 Illustrating the concept of controllability. (**a**) State P2 can be controlled directly by choosing different experimental setups. (**b**) The reaction R1 is controlled by P2, as P2 is a substrate of R1. (**c**) Control is propagated from R1 both *forward* (to its products) and *backward* (to its substrates). (**d**) Applying this propagation iteratively allows us to analyze the remaining model. (**e**) We find that state P2 exerts control over the entire model. (**f**) If the reaction R1 is removed, P2 loses its entire ability to control the model

A state S_n is active:

- If it has an initial concentration larger than zero
- If $\mathrm{Prod}(S_n)$ contains any active reactions

A reaction R_n is active:

- If all states in $(\mathrm{Subs}_S(S_n) \cup \mathrm{Cat}_S(S_n))$ are active

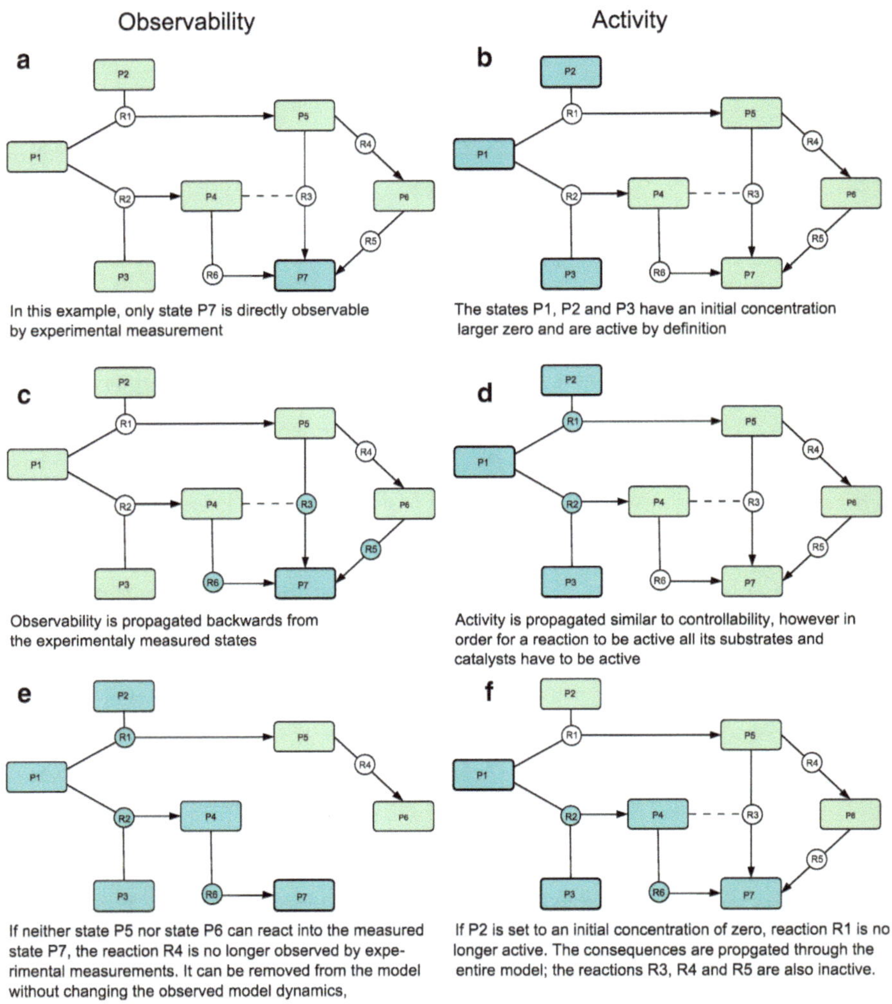

Fig. 3.7 Activity and observability. Activity (*right side*) and observability (*left side*) are analyzed similar to activity; however, some important differences exist. (Observability) In contrast to Controllability, the initially observed states are the model outputs. A reaction is observable if either one of its products or substrates is observable. It will propagate this observability to all its substrates and catalysts (but, in contrast to Controllability, not to its products). (Activity) All states with an initial concentration larger than zero are initially active. Frequently, the set of initially active states will either be identically to or a superset of the controlled states. However, it is insufficient that one substrate or catalyst of a reaction is active to propagate activity to a reaction. Instead, all substrates and catalysts have to be active. Reactions in turn only propagate activity in a forward fashion to all their products, but not to their substrates or catalysts

Summary

Controllability
Input State SI:

Controls itself by definition.
Controls a reaction R_n if $(\text{Subs}_R(R_n) \cup \text{Cat}_R(R_n))$ contains any controlled states, i.e., if any substrate or catalyst of is R_n controlled.
Controls a state S_n if $(\text{Subs}_S(S) \cup \text{Prod}(S_n))$ contains any controlled reactions, i.e., if any reaction that either consumes or produces S_n is controlled.

Observability
A state S_n:

Is observable by definition if S_n is an output state
Is observable if $(\text{Subs}_S(S_n) \cup \text{Cat}_S(S_n))$ contains any observable reaction

A reaction R_n:
Is observable if $(\text{Subs}_R(R_n) \cup \text{Prod}(R_n))$ contains any observable state

Activity
A state S_n:
Is active if it has an initial concentration larger than zero
Is active if $\text{Prod}(S_n)$ contains any active reactions

A reaction R_n:
Is active if all states in $(\text{Subs}_R(R_n) \cup \text{Cat}_R(R_n))$ are active

Modeling Implications
A model that contains any inactive or unobservable reactions is *redundant.*
A model that contains any inactive output state is *invalid.*
A model that violates any experimentally established input/output dependencies is *invalid.*

3.5 The Reduction Graph Data Structure

In order to better visualize the concepts introduced in Sect. 3.3, it can be helpful to further analyze the structure of the reduction space in a way that does not depend on the model we want to reduce. We already characterized the reduction space as the powerset of all reduction candidates. We can utilize this by analyzing the inclusion structure, i.e., the subset/superset relation between its elements. This inclusion structure can be visualized by a special graph called Hasse diagram. Hasse diagrams are utilized to visualize Partially ordered sets, of which powersets are one example.

Any reduction candidate R_1 that is a subset of another candidate R_2 will be considered an ancestor of R_2. If, in addition, R_1 has exactly one element less than R_2, R_1 will be called the direct ancestor or parent. Inversely, R_2 will be called a descendant of R_1 if R_2 is a superset of R_1 and a direct ancestor or child of R_1 if R_2 has exactly one element more than R_1.

To generate a Hasse diagram, we construct a graph in which each reduction candidate (including the empty set) is assigned one node. Every node is connected to its direct descendants by a directed edge. The result is a hierarchical, directed acyclic graph starting from the empty reduction (which has no incoming edges) to the complete reduction (which has no outgoing edges). All elements of one hierarchical level remove exactly the same number of reduction candidates.

The theorem in Sect. 3.3.1, can be reinterpreted in this context:

1. If a reduction is identified as valid, all nodes that have a directed edge leading to this reduction can also be marked as valid reductions
2. If a reduction is identified as invalid, nodes that can be reached from this reduction can also be marked as invalid

Based on this interpretation, we can reinterpret the analysis of the reduction set as a path-finding problem in a graph: we are interested in all nodes that can be reached by directed paths that start at the empty reduction set node. This allows us to utilize *path-finding algorithms* designed for different graph-based problems with little adaptions.

When analyzing the reduction graph of a model, it becomes obvious that most valid and invalid reductions can be verified/rejected indirectly. Based on the considerations in the previous section we find that every valid reduction that also has a valid descendant can be validated indirectly. Likewise, any invalid reduction that has an invalid ancestor can be rejected without the need for explicit model checking.

Only two types of reductions have to be tested explicitly. Invalid reductions that have only valid parents have to be rejected by direct testing. We will call these reductions *minimal invalid reductions*. The description "minimal" is used to clarify that all models that remove only a subset of a minimal invalid reduction are valid. Similarly, valid reductions that have no valid children will be called *maximal valid reductions*. To completely analyze the reduction space of a model, it is both necessary and sufficient to find both the set of *minimal invalid reductions* and the set of *maximal valid reductions*. However, in reality, we find that direct identification of these sets is rarely possible. Instead, our goal is to find heuristic search strategies that minimize the direct testing performed for reductions that are neither maximal valid nor minimal invalid (Fig. 3.8).

3.6 Search Strategies

The strategies we will introduce in this chapter all utilize a candidate list that is updated with each accepted or rejected reduction step. If a reduction is accepted, all its direct descendants that have not already been rejected are added to the end of the candidate list. If a reduction is rejected all its descendants that are currently in the candidate list are removed. The order in which candidates are picked from the list (e.g., oldest first, newest first...) determines the exact algorithm. These strategies are strongly similar to path-searching algorithms for graphs. A common property of all candidate-generating strategies is that every reduction directly tested (except the empty reduction set) will always have at least one valid parent.

3.7 Basic Search Strategies

A *Breadth first search* strategy is one of the basic approaches to analyzing the reduction graph. It is implemented as a candidate generating strategy that always picks the first element of the priority queue, i.e., the oldest element as new candidates are added to the end of the queue. This is also called a first in first out priority strategy.

There are strong similarities between this search strategy and the a priori algorithm for frequent item set mining as introduced by Agrawal et al. (1994). Several results regarding the best- and worst-case runtime of the a priori algorithm can be transferred to the Breadth first search strategy. Like the a priori algorithm, a breadth first search operates in a semideterministic fashion. If multiple breadth first runs are started, they will always explicitly test the same reductions, potentially in a varying order.

Based on the analysis of the a priori algorithm, we can also make observations regarding the number of unnecessary explicit tests performed. A breadth first search will only test nodes whose ancestors have all been verified, i.e., accepting a node during a breadth first search will never result in additional information gain. Vice versa, all nodes that will be rejected will be minimal invalid reductions, as they have no invalid ancestors. The result is that the breadth first search will always explicitly test all valid nodes, but in turn only test those invalid nodes that cannot be avoided to be tested.

Based on this we find two general applications; if a model either has only a very small number of valid reductions or rejecting an invalid reduction is on average significantly more costly than accepting a valid reduction.

A *depth first* search strategy can be implemented very similar to a breadth first search, with the difference that the last item of the candidate queue is picked at each step, i.e., the item that has been added most recently (a last in first out priority strategy). However, the resulting search dynamics will strongly diverge from the behavior of a breadth first algorithm. In general, the performance of a depth first search will vary significantly between different restarts.

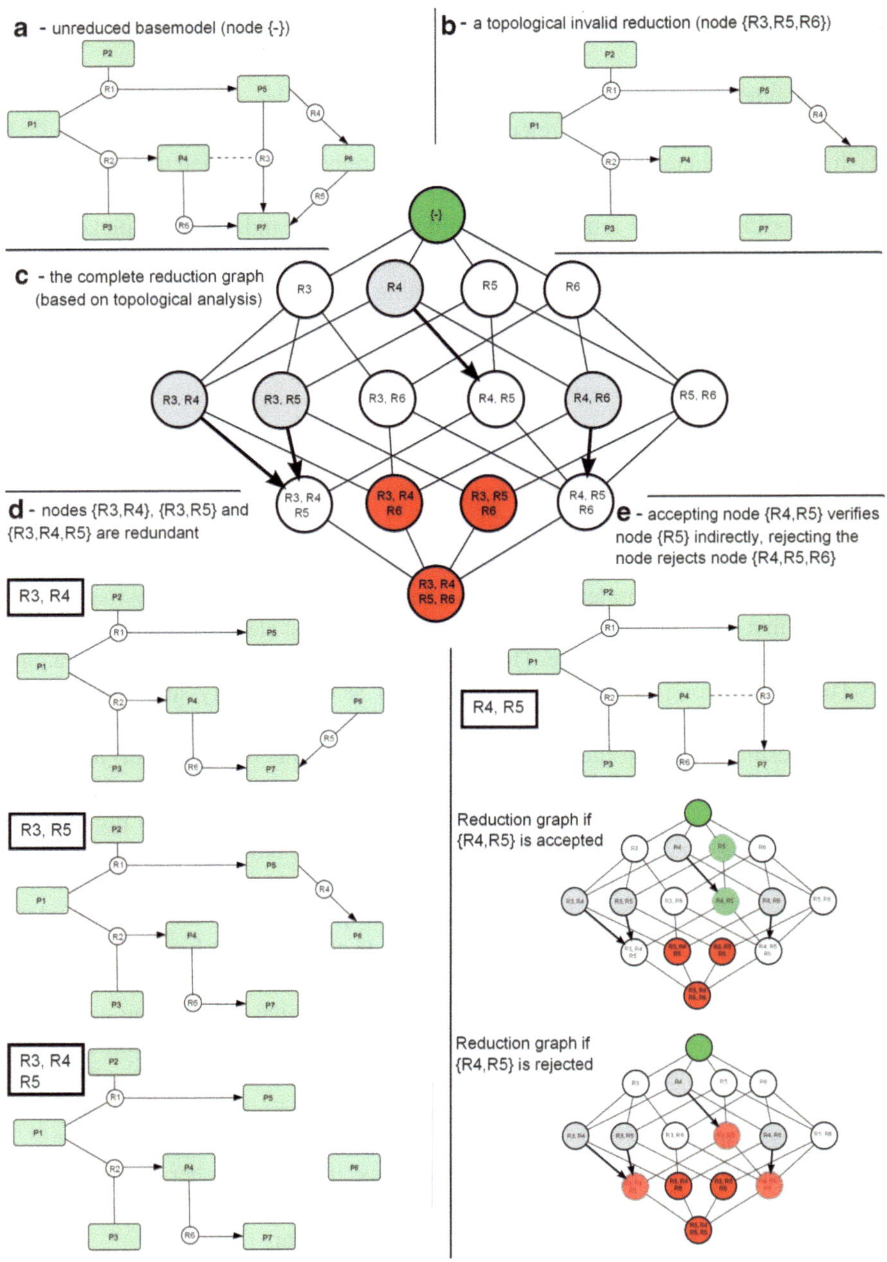

Fig. 3.8 The reduction graph data structure. (**a**) The unreduced base model is mapped to the root node (−) or empty reduction of the reduction graph. It is always valid. (**b**) The reduction B is topologically invalid. It is mapped to node (R3, R5, R6). (**c**) The initial reduction graph for the base model (**a**), after topological analysis but prior to starting a reduction run. Note that only the reactions R3–R6 are reduction candidates, R1 + R2 are considered established reactions. *Red nodes* are topologically invalid, *gray nodes* are redundant. Redundant nodes are connected

A depth first search will most of the time attempt to verify reductions larger than the last accepted reduction. Only if all descendants of the current reduction are rejected will the depth first search start to trace back to earlier reductions. Ideally, the search will find large valid reductions early in its course, indirectly verifying a large number of valid reductions, thus significantly outperforming the Breadth first search. However, if a small invalid reduction is missed, a depth first search can end up getting "stuck" rejecting all its descendants in an unfavorable order. The following example illustrates this problem and compares the approach of the depth first to the breadth first approach.

Both breadth- and depth first approaches will perform very badly if the space of valid reductions is structured in certain ways. The number of verifications required by breadth first approaches grows proportionally to the size of the solution space even if the solution space is structured very regular. In contrast, depth first searches can get stuck in irregular-shaped solution spaces. Therefore it makes sense to include a *random walk*-based strategy as a benchmarking baseline. In such a strategy a random member of the priority queue is chosen at each step. Interestingly, we find that such a random walk-based search will often outperform both breadth- and depth first approaches. This illustrates that the problematic cases discussed for breadth- and depth first approaches occur with significant frequency in modeling applications, and that strategies to deal with these cases are required (Fig. 3.9).

3.8 Hybrid Switching Approach

If we compare the performance of breadth-first, depth-first, and random walk search during an reduction run as illustrated in Fig. 3.10, we find recurring properties. At the start of the reduction run, the depth-first strategy will often outperform both alternative strategies. The depth first strategy initially identifies large valid reductions with a higher frequency then both alternatives. This offers a significant indirect information gain by indirectly verifying a large number of smaller reduction candidates. However, it subsequently gets stuck rejecting a large number of reductions with very small indirect information gained for each rejection.

In contrast, the breadth first search exclusively gains indirect information by rejecting invalid reductions, as has already been discussed. This will often result in

Fig. 3.8 (continued) by an *arrow* to the reduction that is obtained by removing all obsolete elements from them. If a search encounters such a redundant node, it automatically skips to the node indicated by the *arrow* as it is a unique representative of this redundancy group. (**d**) Both (R3, R4) and (R3, R5) are redundant. The unique representative of this redundancy group is the reduction (R3, R4, R5). (**e**) Explicitly testing the reduction (R4, R5) can have two results. If it is accepted, this will indirectly verify (R5). If it is rejected, this will indirectly reject (R3, R4, R5) and (R4, R5, R6). The resulting reduction graph for either case is illustrated

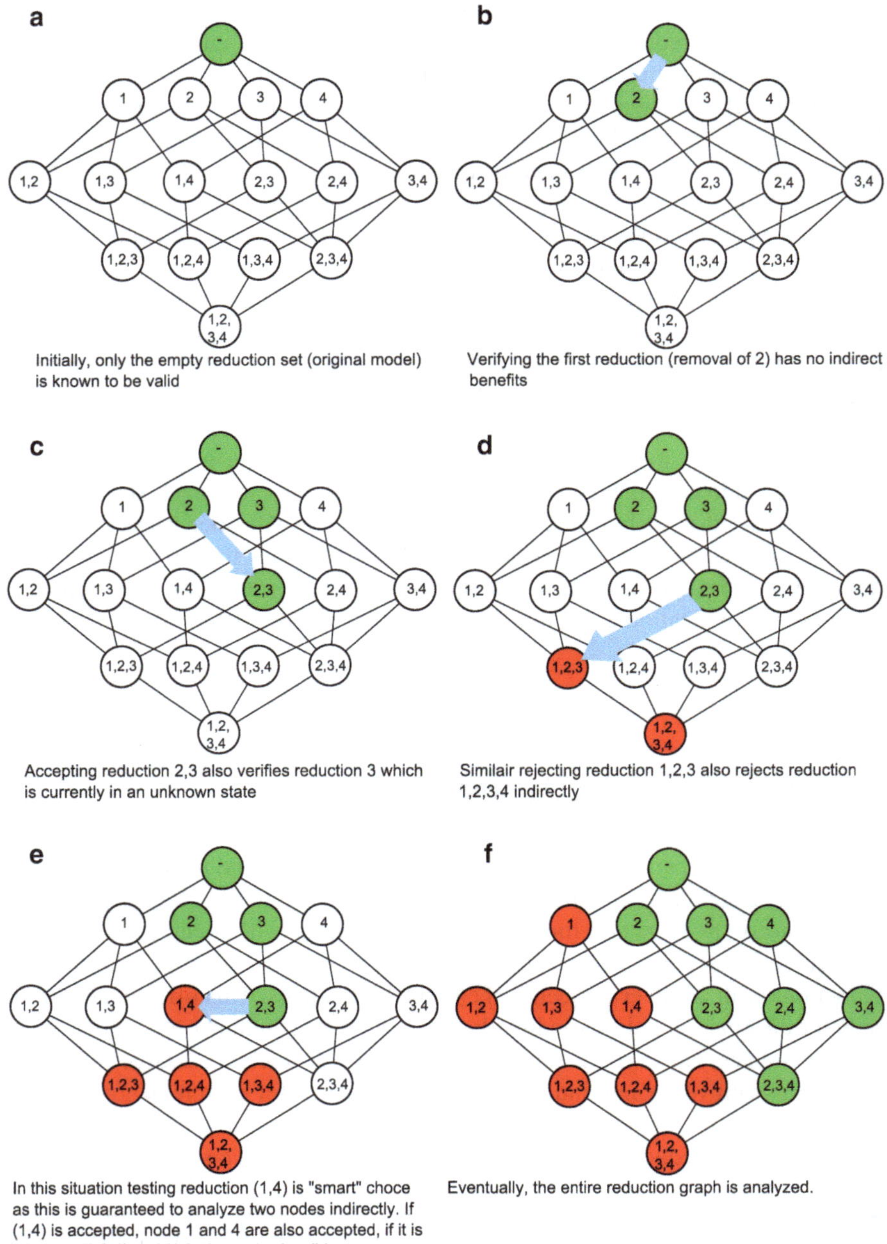

Fig. 3.9 Direct and indirect validation and rejection in the reduction graph. (**a**) Most search strategies will start at the root node of the reduction graph. (**b**) Verifying that (2) is a valid reduction does not provide indirect information gain. (**c**) Verifying node (2,3) also verifies node (3). Note that node (2) has already been verified, so it does not count as indirect information gain, despite being an ancestor of node (2,3). (**d**) Similar to (**c**), rejecting (1,2,3) results in indirect

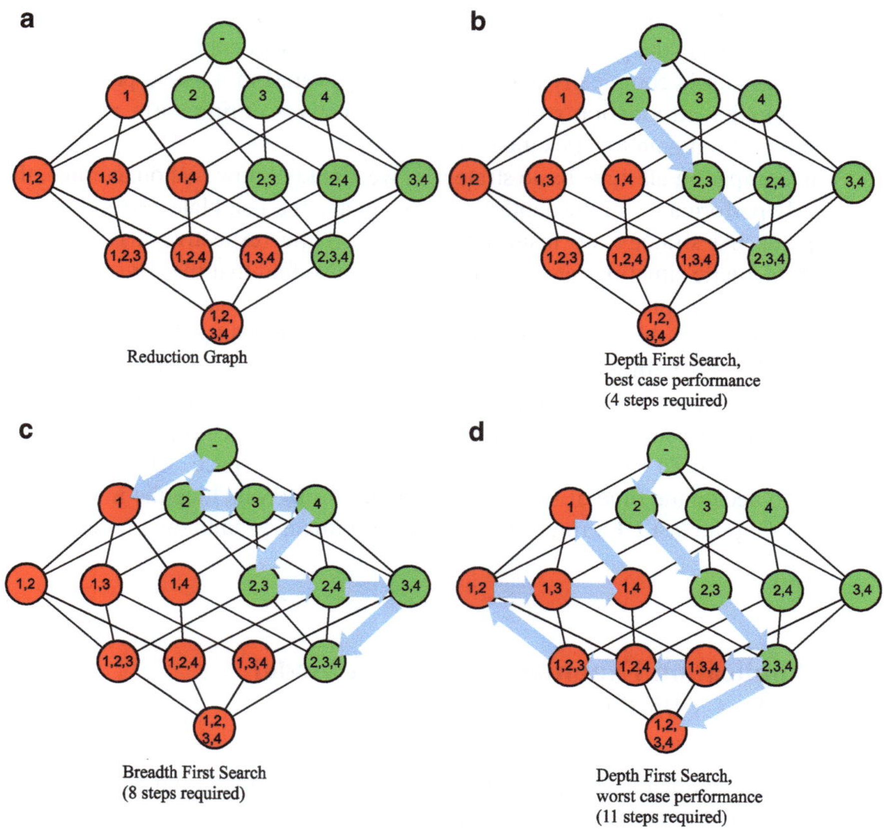

Fig. 3.10 Comparing depth first and breadth first random search. (**a**) Completely analyzed reduction graph. (**b**) Best case of a search trace for a depth first search. The search initially rejects (1) and indirectly rejects all its descendants. It then proceeds to analyze the remaining, valid part of the reduction graph in only four steps. (**c**) A breadth first search will always identify (1) as an invalid reduction during its first few steps. However, it will need to verify all remaining valid reductions explicitly, resulting in a performance that is worse than the best case of a depth first search as illustrated in (**b**). (**d**) However, if a depth-first analysis misses the small invalid reductions and directly wanders to node (2,3,4), the rejections of the invalid reductions will be done by backtracking from a large valid reduction. This case results in the worst-case performance that is worse than the breadth first search

Fig. 3.9 (continued) information gain by indirectly rejecting (1,2,3,4) (**e**) In the situation as displayed in **d**, the analysis of node (1,4) a next step is a smart choice. Either accepting or rejecting it will result in the indirect analysis of two nodes. Note that no parent of (1,4) has been analyzed; therefore, only a candidate picking, but not a candidate generating search strategy would be able to utilize this reasoning to skip to node (1,4). In this example, node (1,4) is rejected, resulting in the additional rejection of (1,2,4) and (1,3,4). (**f**) Once the complete graph is analyzed, we know all valid reductions of the initial model

a slow initial phase, where a large number of small valid reductions are explicitly tested. In exchange, the problem of long stretches of subsequent rejection with little information gain is completely avoided, resulting overall in a stable performance that has smaller information gains then the alternatives during early phases and larger information gain towards the end.

From comparing all three basic strategies, we already know that on average, the good early performance of the depth-first strategy and the good late performance of the breadth-first strategy are insufficient to set off the respective disadvantages when problematic situations are encountered. We also find that the advantages and disadvantages of depth- and breadth-first search supplement each other.

An alternative to the weighted random walk we call hybrid switching approach is to initially start with a DFS that is interrupted as soon as a certain number of rejections has been reached. Once this has happened, either a new DFS is started (that would use a path different from the initial DFS) or the DFS phase is stopped and a BFS is started to analyze the remaining reduction space. Criteria that are possible to decide the time of switching from DFS to BFS include the information gained during the last DFS run or the number of remaining unknown reductions.

3.9 Application Example: Reducing a Model of the CD95 Pathway

Regulation of cell death decisions via CD95 signaling involves complex dynamics of the involved pro- and anti-apoptotic proteins, e.g., procaspase-8 and c-FLIP, and their cleavage products. These sometimes interact in surprising, non-intuitive ways. A signal that induces cell proliferation and survival at low concentrations can induce apoptosis at higher concentrations, thus resulting in opposing effects depending on whether a threshold is met (Lavrik et al. 2007). To understand not only the qualitative level of these regulatory mechanisms, but the details of the molecular interactions resulting in such a threshold behavior, researchers have begun developing quantitative signaling models (Fricker et al. 2010). While these models are currently able to illustrate the molecular dynamics encountered, they typically suffer from indeterminacies stemming from either over specified models or biologically relevant alternative architectures. This reduces their value in model-based prediction, as parameter uncertainties will often directly result in uncertain and ambiguous predictions. In this section, we will illustrate how model reduction can be utilized to improve our understanding of the processes happening during this signaling and to derive new models that better represent these processes.

To analyze the role of c-FLIP cleavage in apoptosis induction, we have used a model describing the apoptotic branch of the CD95 signaling pathway as implemented in (Fricker et al. 2010). This model is illustrated in Fig. 3.11. While the model explains the interaction of c-FLIP in an intuitive way, it is considerably underdetermined. Many intermediate states of the pathway can only be measured as

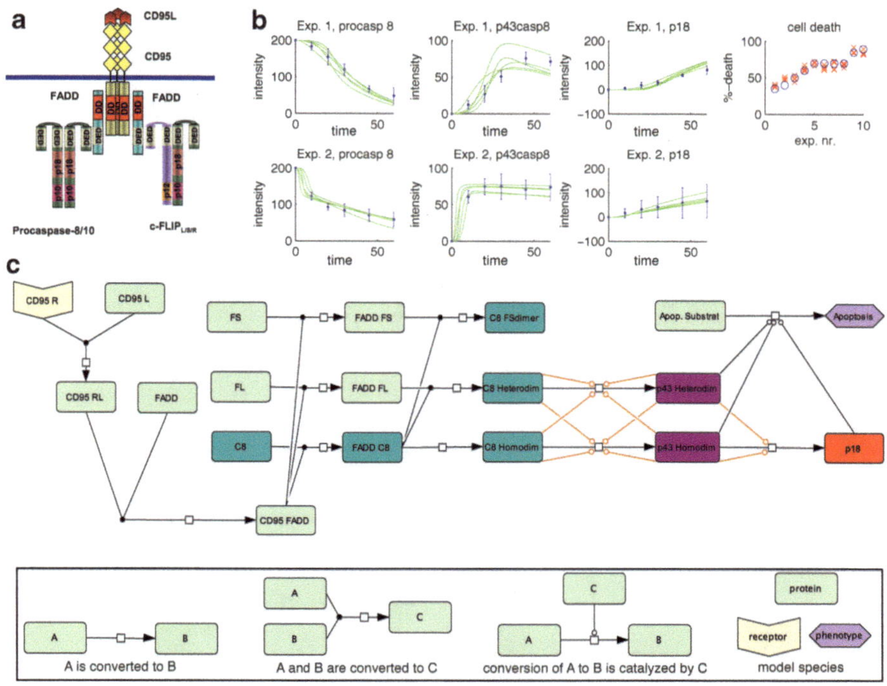

Fig. 3.11 Biological background, experimental data, and computational model of the CD95 pathway. (**a**) The DISC (Death Inducing Signaling Complex) is formed at the membrane. Its most important parts are the CD95 ligand/receptor complex and FADD (Fas Associated Death Domain) protein. Both Procaspase-8/10 and c-FLIP$_{S/L}$ are recruited by and bound to this complex and further processed and activated. (**b**) Experimental measurements of various molecules in the CD95 pathway during stimulation experiments. For protein measurements, the x-axis denotes time while the y-axis denotes relative intensity. *Blue error bars* are the measured points, *green lines* are simulations by the pathway model (**c**) with different parameter sets. All simulations have similar overall quality. For the cell death measurements, the x-axis denotes the number of the experiments, each *blue circle* is a separate experiment. The y-axis denotes the number of cells that had died by the end of the experiment. Red x's are again the result of simulating the CD95 model with different parameter sets. (**c**) The CD95 model. Equally colored species were measured as one experimental concentration; dissociation reactions were omitted for clarity. Reactions that are candidates for removal have been highlighted in red. Our goal is to find model reductions that are roughly as good as the simulations illustrated in (**b**), but contain fewer of the reactions marked as reduction candidates

groups and a significant number of potential reactions cannot be observed directly as they occur in membrane-localized complexes that are difficult to measure in an experimental context.

The model is roughly divided into three parts. The first part of the model simulates the binding of the extracellular CD95L (CD95 ligand) to CD95R (CD95 receptor) (Suda et al. 1993). The activated receptor recruits FADD. Bound FADD multimerizes and thus creates the membrane-localized DISC (Death Inducing Signaling Complex), denoted as CD95 FADD in the model (Kischkel et al. 1995).

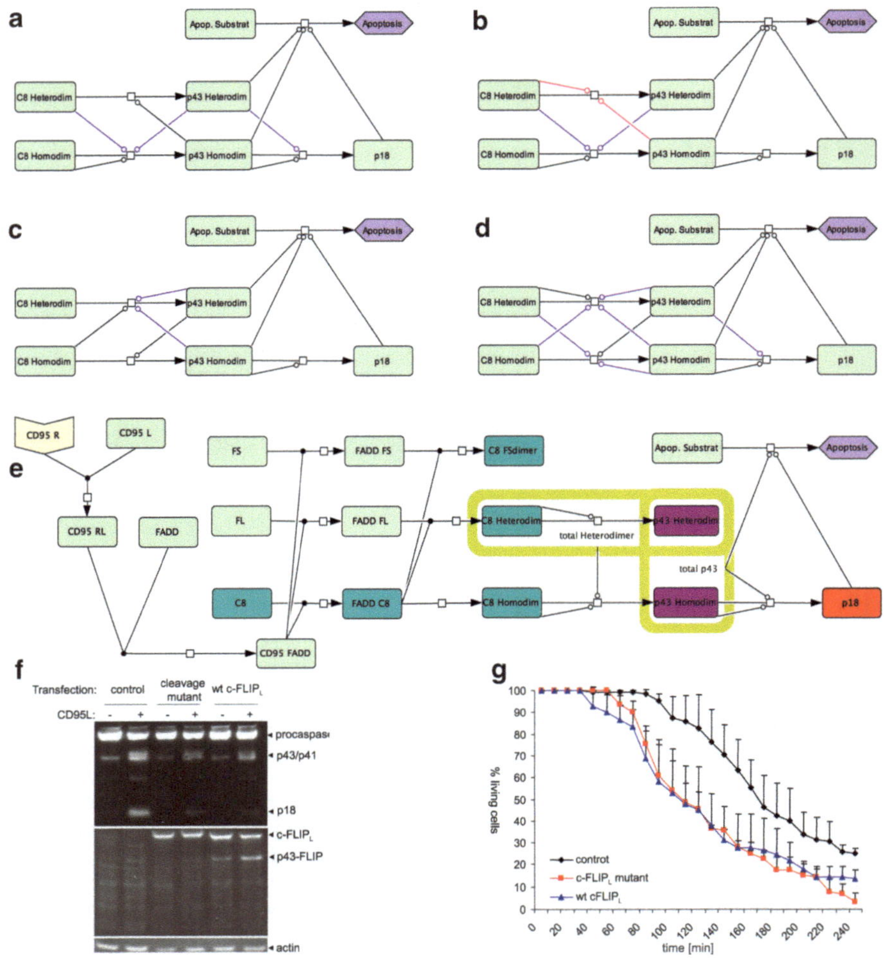

Fig. 3.12 Model reductions of the full CD95 signaling model and CD95 consensus model. The 14 reduced models are summarized in (**a–d**) by including any one (**c**, **d**) or two (**a**) edges of the same color or one edge of each color (**b**). A recurring pattern is that all reductions include the activation of the C8 homodimer by either the C8 or the p43 heterodimer. This illustrates that clearly the interaction of homo- and heterodimers is an essential part of the pathway dynamics and that limiting the model to homo–homo and hetero–hetero interactions is not a valid approach. The minimal reductions summarized in (**c**) can be considered questionable in a biological context; it seems unlikely that the C8 homodimer can activate the C8 heterodimer but not itself while the C8 heterodimer cannot act as catalyst at all. This illustrates that our current set of experimental data is insufficient to completely characterize the CD95 pathway in a satisfying fashion. The same holds true for the reductions summarized in (**d**); it seems biologically questionable that C8 homodimers show no catalytic activity. Based on these considerations we derive a new consensus model based on the minimal reductions summarized in (**a**) + (**b**). Both (**a**) and (**b**) retain the autocatalytic activation of the C8 homodimer. In addition, both models have to include the activation of the C8 homodimer by either the C8 or the p43 heterodimer. As either reaction is fine, it can be reasoned that current data suggests that both reactions are dynamically very similar. The resulting hypothesis that the C8 homodimer is activated by both the C8 heterodimer and the p43 heterodimer with

The second part of the model summarizes the binding reactions of procaspase-8 and c-FLIP isoforms to the DISC which results in the formation of different types of dimers. Three types of catalytically active dimers are formed: procaspase-8-homodimers (called C8 dimers), procaspase-8/c-FLIP$_L$ heterodimers (called C8 heterodimers), and procaspase-8/c-FLIP$_S$ heterodimers (Neumann et al. 2010).

The third part of the model focuses on the activation of the different procaspase-8 dimers. C8 dimers and C8 heterodimers are proteolytically activated by (spatially) neighboring dimers and further processed. In contrast the procaspase-8/c-FLIP$_S$ heterodimer is not processed further. The procaspase-8 part of the dimers is cleaved into the active form p43/p41, resulting in p43/p41 homo- and heterodimers. p43 homodimers are further processed into the caspase-8 tetramer containing the cleavage product p18.

The proteolytically activated heterodimers p43/p41 and p18 forms of procaspase-8 can contribute to the progression of apoptosis by activating various downstream effector caspases. This effect is summarized as the cleavage of apoptosis substrate. The model state "apoptosis" is measured by rate of cell death experimentally.

The part of the model that is most severely underdetermined is the activation of both C8 homodimer and C8 heterodimer. It is known that this reaction has to be catalyzed; however, C8 homodimer, C8 heterodimer, p43 homodimer, and p43 heterodimer are all candidates as possible catalysts for both reactions. The initial model therefore includes four activating reactions for each C8 homodimer and C8 heterodimer, one for each possible catalyst. A similar situation occurs for the activation of apoptosis substrate; here, both p43 heterodimer and p43 homodimer are potential candidates for catalyzing this reaction.

We applied the model reduction approach as introduced in the previous chapter to the unreduced base model of the CD95 pathway, using all reactions that activated either the C8 homodimer, the C8 heterodimer, or the apoptosis substrate as reduction candidates. The total set of reduction candidates contained 11 reactions, resulting in $2^{11} = 2{,}048$ possible reduction sets. Redundancy and validity analysis reduced the unknown model space by about 45% to a total of 1,158 nonredundant reductions.

Running the reduction search identified 237 reductions as valid, the 921 other reductions were invalidated. The model space can be characterized by 14 minimal valid reductions and eight maximal invalid reductions, as illustrated in Fig. 3.12. All minimal models reproduce the experimental data with an error score comparable to the unreduced base model. All valid reductions included the activation of

Fig. 3.12 (continued) the same rate is consistent with modeling results. Additional experiments with a c-FLIP$_L$ cleavage mutant were performed and compared to the behavior of wildtype c-FLIP$_L$ (**f**). In the mutant cell line the activation of C8 heterodimer to p43 heterodimer is blocked, no p43 heterodimer is produced. Quantification of cell survival (**g**) shows that cell survival rates are the same for wildtype and cleavage mutant cells, supporting the hypothesis that the C8 heterodimer activation does not change its catalytic influence on C8 homodimer activation.

apoptosis substrate by the p43 heterodimer. The largest valid reduction removes six reactions, any reduction removing seven reactions is invalid.

Every valid reduction includes the activation of the C8 homodimer by either the C8 heterodimer or the p43 heterodimer. In turn the autocatalytic activation of the C8 homodimer or the activation of the C8 homodimer by the p43 heterodimer is also included. The minimal satisfactory solution for these two conditions, the activation of C8 homodimer by the p43 heterodimer, but not by either itself or the C8 homodimer is part of various valid reductions. Indeed, the autocatalytic activation of the C8 homodimer is retained only in 4 out of 14 minimal reductions, although there is a strong biochemical evidence reported showing formation of C8 homodimers at the DISC.

The interpretation of these properties confirms that the role of c-FLIP$_L$ cannot simply be reduced to that of an inhibitor of CD95 signaling. Instead, the C8 heterodimer acts as a catalyst for the activation of C8 homodimers, either directly or in the activated p43 heterodimer version. However, various reported mechanisms of caspase-8 activation cannot be verified based on the experimental data currently available. This mainly concerns the catalytic activity of C8 both in homo- and heterodimer form. We expect that the autocatalytic activation of C8 homodimers is an essential process in the CD95 apoptosis signaling; yet our experimental data does not reflect this.

References

Agrawal R, and Srikant R Fast algorithms for mining association rules in large databases. Proceedings of the 20th International Conference on Very Large Data Bases, VLDB, pages 487–499, Santiago, Chile, September 1994.

Fricker N, Beaudouin J, Richter P, Eils R, Krammer PH et al (2010) Model-based dision of CD95 signaling dynamics reveals both a pro- and antiapoptotic role of c-FLIPL. J Cell Biol 190 (3):377–389, Intro to CD95 Modeling

Kischkel FC, Hellbardt S, Behrmann I, Germer M, Pawlita M et al (1995) Cytotoxicity-dependent APO-1 (Fas/CD95)-associated proteins form a death-inducing signaling complex (DISC) with the receptor. EMBO J 14(22):5579–5588, Forming of DISC

Lavrik IN, Golks A, Riess D, Bentele M, Eils R et al (2007) Analysis of CD95 threshold signaling: triggering of CD95 (FAS/APO-1) at low concentrations primarily results in survival signaling. J Biol Chem 282(18):13664–13671, Intro to threshold

Neumann L, Pforr C, Beaudouin J, Pappa A, Fricker N et al (2010) Dynamics within the CD95 death-inducing signaling complex decide life and death of cells. Mol Syst Biol 6:352, Role of cFLIPl

Suda T, Takahashi T, Golstein P, Nagata S (1993) Molecular cloning and expression of the Fas ligand, a novel member of the tumor necrosis factor family. Cell 75:1169–1178, CD95R/L Binding

Systems Biology Toolbox for MATLAB: A computational platform for research in Systems Biology, Bioinformatics, 22(4):514–515, 2006.

Chapter 4
Systems Biology of the Mitochondrial Apoptosis Pathway

Jochen H.M. Prehn, Heinrich J. Huber, and Carla O'Connor

Abstract Mitochondria have multiple functions. Apart from their role in the regulation of cellular bioenergetics, redox homeostasis and signal transduction, mitochondria are able to initiate apoptosis. The Bcl-2 family proteins are the key regulators of the mitochondria-initiated caspase activation pathway. Activation of caspases is considered one of the most important regulatory steps for apoptosis. Caspase cascades can be initiated or amplified by the release of cytochrome-*c* from the mitochondria. On release into the cytosol, cytochrome-*c* binds to Apaf-1. Apaf-1 oligomerises and engages the initiator caspase, pro-caspase-9, which in turn activates downstream caspases-3 and -7. Furthermore, the release of Smac from mitochondria assists the amplification of the caspase cascade by abrogating the function of caspase inhibitors such as XIAP. Moreover, mitochondria are involved in caspase independent cell death with the release of factors including apoptosis-inducing factor. There have been considerable developments in recent years in further understanding the complex signaling networks and cellular decision making during mitochondria-initiated apoptosis through the use of systems biology. In this review, we examine the modeling approaches that are currently employed to further our understanding of the mitochondrial pathway of apoptosis.

J.H.M. Prehn (✉) • H.J. Huber
Department of Physiology and Medical Physics, Centre for Systems Medicine
Royal College of Surgeons in Ireland, 123 St. Stephen's Green, Dublin 2, Ireland
e-mail: prehn@rcsi.ie

C. O'Connor
Department of Physiology and Medical Physics, Centre for Human-Systems Medicine,
Royal College of Surgeons in Ireland,
123 St. Stephen's Green, Dublin 2, Ireland

I.N. Lavrik (ed.), *Systems Biology of Apoptosis*, DOI 10.1007/978-1-4614-4009-3_4,

Abbreviations

CTL	Cytotoxic T-lymphocytes
NK	Natural Killer
Apaf-1	Adaptor protein, apoptotic protease activating factor-1
Bcl-2	B-cell lymphoma protein 2
IAP	Inhibitor of Apoptosis
MOMP	Mitochondrial Outer Membrane Permeabilisation
BH	Bcl-2 homolgy
BIR	Baculoviral IAP Repeat
RING	Really Interesting New Gene
c-IAP1 and c-IAP2	Cellular inhibitor of apoptosis 1 and 2
NAIP	Neuronal Apoptosis Inhibitor Protein
XIAP	X-linked inhibitor of apoptosis protein
ILP-2	Inhibitor of Apoptosis-Like Protein-2
Smac	Second Mitochondria-derived Activator of Caspases
DIABLO	Direct IAP Binding Protein with Low PI
ODE	Ordinary Differential Equations
BAR	Bifunctional Apoptosis Regulator
NF-κB	Nuclear Factor-KappaB
IKK	IκB kinase
FRET	Foerster resonance energy transfer
TRAIL	Tumour Necrosis factor-Related Apoptosis-Inducing Ligand
CHX	Cycloheximide
ATP	Adenosine Triphosphate
CA	Cellular Automata

4.1 "Mitochondrial "or "Intrinsic" Apoptosis Pathway

In mammalian cells there are at least three main pathways which lead to caspase activation, the intrinsic or mitochondrial pathway, the extrinsic or death receptor pathway and the cytotoxic T-lymphocytes (CTL)/natural killer (NK)-derived granzyme B-dependent pathway. There is a certain amount of crosstalk between the pathways and all may ultimately result in the apoptotic death of the cell.

The intrinsic pathway of apoptosis can be initiated by various forms of stress such as DNA damage, trophic factor withdrawal, nutrient deprivation, heat shock and oxidative stress. Apoptotic signaling in this pathway results in an increase in the permeability of the mitochondrial outer membrane and the subsequent release of several proteins from the inter-membrane space of the mitochondria into the cytoplasm, resulting in the activation of both initiator and effector caspases. Central to this pathway is the release of cytochrome *c* from the mitochondria (Kluck et al. 1997a, b; Liu et al. 1996). Normally cytochrome-*c* resides in the inter-membrane space of the mitochondria, where it functions by transporting electrons between protein complexes of the respiratory chain during oxidative phosphorylation. On release of cytochrome-*c* into the cytosol, it binds to and activates the adaptor protein, apoptotic protease activating factor-1 (Apaf-1) (Li et al. 1997; Zou et al. 1997).

Oligomerisation of Apaf-1 results in the recruitment of the initiator caspase pro-caspase-9 and the formation of the apoptosome (Acehan et al. 2002; Shi 2006). Once activated, the mature caspase-9 remains part of the apoptosome complex, with Apaf-1 functioning as an allosteric regulator of caspase-9 activity, allowing it to cleave and activate downstream effector caspases such as pro-caspase-3 and pro-caspase-7 (Adrain et al. 1999; Rodriguez and Lazebnik 1999; Slee et al. 1999). The executioner caspases-3 and -7 exist within the cytosol as inactive dimers (Boatright et al. 2003; Donepudi and Grutter 2002). When activated, these caspases cleave and activate further downstream caspases such as caspases-2 and -6. Caspase-3 is also involved in a feedback amplification loop to further activate caspase-9 (Slee et al. 1999).

4.2 Regulators of the Mitochondrial Apoptosis Pathway

The importance of properly balanced cell survival and death in an organism is undeniable. Unscheduled survival and proliferation of cells beyond their natural life span can lead to the formation of tumours and cancer, while, at the other end of the spectrum, the premature death of differentiated cells such as neurons or cardiac muscle cells leads to irreversible, degenerative diseases. Given the complexity of the signaling involved, it is not surprising that a large variety of endogenous regulators of apoptotic signaling have been identified and investigated in mammalian cells. Of particular importance for the mitochondrial apoptosis pathway are the B-cell lymphoma protein 2 (Bcl-2) and inhibitor of apoptosis (IAP) protein families.

4.2.1 The Bcl-2 Family

The Bcl-2 family of proteins are probably the best described endogenous modulators of the mitochondrial pathway of apoptosis, and regulate apoptosis by either promoting or preventing mitochondrial outer membrane permeabilisation (MOMP) (Hengartner and Horvitz 1994; Adams and Cory 2007). Pro- and anti-apoptotic family members can heterodimerise and neutralise each other's function. The Bcl-2 family proteins can be divided into three subfamilies. The first subfamily comprises the anti-apoptotic proteins (Bcl-2, Bcl-x_L, Bcl-w, Mcl-l, Al and Boo) (Hockenbery et al. 1990; Oltvai et al. 1993) which are potent inhibitors of the apoptotic programme and antagonise pore formation at the mitochondrial outer membrane. These proteins are structurally characterised by four Bcl-2 homolgy

(BH) domains. The remaining two subgroups, the Bax/Bak family and the BH3-only protein family, are pro-apoptotic and required for the initiation of apoptosis. The multi-domain, pro-apoptotic proteins, Bax and Bak (and potentially a third protein, Bok), bear three BH domains, and oligomerise and facilitate pore formation in the outer mitochondrial membrane. The BH3-only proteins promote apoptosis by not only selectively binding to the anti-apoptotic Bcl-2 family members, but also by directly inducing the activation and oligomerisation of the Bax and Bak proteins (Adams and Cory 2007; Danial and Korsmeyer 2004; Youle and Strasser 2008). Activation of Bax and Bak is essential for the activation of the mitochondrial apoptosis pathway (Wei et al. 2001).

It should also be noted that there is crosstalk between the mitochondrial and death receptor apoptosis pathways. This crosstalk is carried out by caspase-8-mediated cleavage of the BH3-only protein family member, Bid (Li et al. 1998; Luo et al. 1998).

4.2.2 Inhibitor of Apoptosis Proteins

The Inhibitor of Apoptosis Proteins (IAPs) regulate apoptosis by direct inhibition of caspases by preventing their cleavage (Deveraux and Reed 1999; Deveraux et al. 1998; Riedl et al. 2001; Shiozaki et al. 2003). Interestingly, IAPs seem to be multifunctional and are not only involved in regulating apoptosis, but are also involved in cell signaling, inflammation and cell cycle progression (Hofer-Warbinek et al. 2000; Huang et al. 2000; Lu et al. 2007; MacFarlane et al. 2002; Sanna et al. 1998; Varfolomeev et al. 2007). All IAPs contain one to three Baculoviral IAP repeat (BIR) motifs, the presence of at least one BIR domain is essential for the anti-apoptotic activity of members of the IAP family (Vucic et al. 1998). Several, but not all, IAPs also contain a carboxy-terminal Really Interesting New Gene (RING) zinc finger, which has been recognised to have E3 ubiquitin ligase activity (Yang et al. 2000). To date eight members of the IAP family have been identified in humans. These include cellular inhibitor of apoptosis 1 and 2 (c-IAP1) and (c-IAP2), Neuronal Apoptosis Inhibitor Protein (NAIP), Survivin, X-linked inhibitor of apoptosis protein (XIAP), Bruce, Inhibitor of Apoptosis-Like Protein-2 (ILP-2), Livin and Apollon (Liston et al. 2003).

4.2.3 XIAP and Smac/DIABLO

Among the IAPs, XIAP is the most potent inhibitor of cell death, capable of blocking both the intrinsic and extrinsic pathways of apoptosis through inhibition of the initiator caspase-9 and effector caspases-3 and -7 (Deveraux et al. 1997; Chai et al. 2001; Srinivasula et al. 2001). In addition to its caspase inhibition abilities, XIAP also has a RING motif which functions as an E3 ubiquitin protein ligase to catalyse the ubiquitination of itself as well as substrate proteins (Huang et al. 2000; Yang et al. 2000; Salvesen and Duckett 2002). Caspase inhibition by XIAP may be counteracted by the release of Second Mitochondria-derived Activator of Caspases/ Direct IAP Binding Protein with Low PI (Smac/DIABLO) or the serine protease Omi/HtrA2, both of which are released by mitochondria into the cytosol during apoptosis (Verhagen et al. 2000; Martins et al. 2002).

4.3 ODE-Based Models of the Mitochondrial Apoptosis Pathway

Mathematical models based on ordinary differential equations (ODE) have decisively contributed to apoptosis research by giving a comprehensive insight into previously inexplicable phenomena. These phenomena include the rate of permeabilisation of the mitochondrial membrane (MOMP) and subsequent speed of effector caspase activation, the phenomena of sub-lethal caspase activation and the shift between direct effector caspase activation by the death receptors ("type 1" apoptosis) to the engagement of the Bid-dependent, mitochondrial amplification loop ("type 2" apoptosis) during the extrinsic pathway of apoptosis.

4.3.1 Caspase Activation Models

A theoretical study by Fussenegger and colleagues investigating the activation of the stress-induced mitochondrial apoptotic pathway and the induction of extrinsic pathway by death receptors first introduced the concept of ODE-based modeling to the field of apoptosis (Fussenegger et al. 2000). They reported that the release of cytochrome-*c* as a result of MOMP acted as a determinant of stress-induced cell death between pro-survival and pro-apoptotic cell death. The study predicted a temporal profile of caspase activation which was dependent on the intracellular levels of Bcl-2. In a subsequent analysis by Eissing et al. which focused on modeling of the extrinsic pathway of apoptosis, excluding the involvement of the mitochondria, the authors remodelled the fast kinetics of caspase-3 activation that had been previously reported in single cell studies (Eissing et al. 2004). In further agreement with this model, Pace et al. were able to compare Eissing's model predictions to their in vitro experiments, confirming that reducing the levels of the pro-caspase-8 inhibitor, bifunctional apoptosis regulator (BAR) leads to accelerated caspase-3 activation (Pace et al. 2010). The effect of mitochondrial cytochrome-*c* and Smac release on caspase activation was first modelled by Stucki and Simon. In this study the extrinsic pathway was modelled on the activation of death receptors which resulted in a dose-dependent release of cytochrome-*c*, leading to an increased concentration of caspase-3 (Stucki and Simon 2005).

This increase of caspase-3 was dependent on the level of XIAP, the inhibitor of the caspases-3, -7 and -9. In comparison, the intrinsic pathway was modelled to activate a dose-dependent Smac release, which was antagonised by Survivin, a proposed inhibitor of Smac. Therefore the modeling approach could achieve stability against inadvertent caspase-3 activation as both Smac and caspase-3 were constrained by inhibitors of apoptosis. However as the release of cytochrome-*c* (Goldstein et al. 2000) and Smac (Rehm et al. 2003) has been shown to happen as a rapid, single-step process, with little dependence on the type or dose of stimulus (Huber et al. 2009), and as cytochrome-*c* only binds to the apoptosome transiently (Zou et al. 2003), it seems improbable that the amount of released cytochrome-*c* is rate limiting for caspase-3 activation.

To gain a better understanding of the of CD95 extrinsic apoptosis pathway, Bentele, Lavrik and colleagues subsequently devised an ODE-based mathematical model. In combination with experimental data, the model provided a comprehensive analysis of CD95-mediated apoptosis and determined a threshold mechanism for the regulation of the extrinsic pathway within the cell (Bentele et al. 2004; Lavrik et al. 2007). Modeling, however, also included aspects of the mitochondrial apoptosis pathway. CD95/Apo-1-induced capase-8 activity resulted in the concentration of tBid exceeding the threshold concentration and activating the mitochondrial cell death cascade. Furthermore, this model has been extended to investigate the role of nuclear factor-KappaB (NF-κB) activation in CD95-dependent apoptosis, which has suggested that p43-FLIP acts as an activator of IκB kinase (IKK) (Neumann et al. 2010). The authors were able to demonstrate a temporal association between CD95 stimulation and IKK activity through quantification of Western blot analysis following anti-CD95 stimulation. This important study suggested that a cleavage product of c-FLIP$_L$, p43-FLIP, interacts with the IKK complex and triggers NF-κB activation.

A further example of the use and importance of ODE approach in apoptosis research was shown by Rehm and co-workers, who provided further insights into the mitochondrial signaling cascade by combining single cell imaging of caspase activation with ODE-based modeling (Rehm et al. 2006). The release of the mitochondrial proteins, cytochrome-*c*, and Smac was considered trigerring of effector caspase activation by the model. The model was verified by single cell experiments using a caspase-3/-7-specific Foerster resonance energy transfer (FRET) probe. The authors detected the onset of mitochondrial depolarisation, an indicator of MOMP, which was followed by a rapid onset and cleavage of a caspase-3-specific FRET probe. The ODE-based approach was effectively able to remodel the FRET probe cleavage kinetics of HeLa cells and the influence of caspase-3 deletion or over-expression of XIAP when exposed to pro-apoptotic stimuli. The authors predicted and also demonstrated biological scenarios of incomplete cleavage of cellular substrates. Consequently, this system model was employed once more by O'Connor and co-workers, and provided an understanding into the faster onset of apoptosis observed in XIAP-deficient colon cancer cells (O'Connor et al. 2008).

4.3.2 MOMP Models

The research presented above mainly concentrated on the events downstream of MOMP such as the release and translocation of cytochrome-*c* and Smac on caspase activation; however, the actual mechanism and mathematical modeling of MOMP formation itself remained more intricate. MOMP is controlled by the Bcl-2 family of proteins which contain both anti- and pro-apoptotic proteins, and thus they play a key role in determining cell fate. In order to investigate the exact mechanism of MOMP, Albeck and co-workers developed a mathematical model of extrinsic apoptosis, which was validated experimentally by live-cell imaging of caspase activation and MOMP, flow cytometry and immunoblotting of HeLa cells exposed to the death receptor agonist tumour necrosis factor-related apoptosis-inducing ligand (TRAIL) (Albeck et al. 2008a, b). MOMP induction was integrated through initiator caspase cleavage of Bid which subsequently activated pro-apoptotic Bcl-2 proteins (Bax and Bak), resulting in Bax translocation and Bax- and Bak-induced pore formation. Within the model anti-apoptotic Bcl-2 proteins were deemed as one variable, and the kinetics of Bid cleavage were deducted from kinetics of caspase-8 activation. When MOMP was blocked by Bcl-2 over-expression, the model rendered a meta-stable state, with initiator, but not effector caspase activity. This meta-stable state arose as a result of XIAP inhibition of effector caspases, in addition to proteasome degradation of the caspases, which was further enhanced by the E3 ligase activity of XIAP. In contrast, this stabilisation was abrogated and mild effector caspase activity was predicted when low levels of XIAP were assumed or the degradation of effector caspases was inhibited. This suggested that a non-committal to apoptosis would occur if cleavage of cellular substrates was submaximal. The model's predictions of the existence of a "meta-stable" state were confirmed on using Bcl-2 over-expressing cells that are resistant to MOMP, on treatment with TRAIL either in the presence of the proteasome inhibitor MG132 or by XIAP silencing. Furthermore, the model predicted that a reduction of Bcl-2 levels or increased BH3-only protein levels resulted in the restoration of MOMP and hence a surge in caspase activation.

An additional theoretical model to investigate the role of Bcl-2-family members to induce MOMP and subsequent caspase activation upon intrinsic stress was developed by Zhang and colleagues (Zhang et al. 2009). Their approach consisted of apoptosis initiation, amplification and execution modules that together guaranteed cellular robustness to subthreshold stimuli. In agreement with other models (Albeck et al. 2008a; Chen et al. 2007a, b; Dussmann et al. 2010), the authors did not distinguish between individual pro- and anti-apoptotic Bcl-2 family proteins. As genotoxic stress (Brown and Wouters 1999; Kohler et al. 2008), serum deprivation (Kaufman 1999), or receptor-mediated stress (Kelley and Ashkenazi 2004) each activate a unique subset of BH3-only proteins which specifically interact with their pro-survival counterparts (Chen et al. 2005), a model extension to investigate this stress-specific regulations would be a desirable next step.

4.4 Modeling of Mitochondrial Bioenergetics During Apoptosis

In addition to its role in caspase-dependent cell death, cytochrome-*c* is also an essential component of the mitochondrial respiration chain. Thus, loss of cytochrome-*c* in the mitochondria during MOMP leads to impairment of mitochondrial respiration, adenosine triphosphate (ATP) depletion and depolarisation of the mitochondrial membrane potential $\Delta\Psi m$, which may lead to failure of ionic homeostasis and cell death. Huber and co-workers highlighted the importance of combining ODE models of apoptotic signaling with cellular bioenergetics for studying cell death processes subsequent to MOMP (Huber et al. 2011). Their ODE-based model integrated the molecular mechanisms of mitochondrial respiration and ATP production with caspase-dependent signaling downstream of MOMP. The model investigated the response of HeLa cells to cytochrome-*c* release during apoptosis and bioenergetic disruption. Model predictions were validated by single cell fluorescence microscopy experiments of HeLa cervical cancer cells where MOMP was induced by staurosporine. They first confirmed previous experimental findings that caspase-3 cleavage of respiratory complex I and II was able to exacerbate $\Delta\Psi m$ depolarisation (Ricci et al. 2003). Subsequently, the model identified that increased glucose metabolism subsequent to cytochrome-*c* release could lead to a significant $\Delta\Psi m$ repolarisation, and therefore induce a bioenergetic recovery. Using single cell microscopy, they verified that HeLa cells which had depolarised $\Delta\Psi m$ after MOMP were able to partially repolarise $\Delta\Psi m$ when additional glucose was added to the medium. Since a similar repolarisation was not observed in non-transformed colon epithelial CRL1807 cells, the theoretically identified and experimentally confirmed recovery mechanism may be cancer cell specific, suggesting that high glucose levels may give cancer cells a competitive advantage under conditions of therapeutically induced MOMP.

4.5 Studying MOMP by Cellular Automata

An alternative approach referred to as Cellular Automata (CA) was engaged to model the mechanism of MOMP and Bax/Bak pore formation. In comparison to ODE-based mathematical models, CA models provide us with a modeling approach which is capable of explaining alterations of proteins at distinct locations and time points. CA has been employed to investigate and quantify the relationship of Bcl-2 proteins at the mitochondrial membrane during intrinsic apoptosis. CA is used to study signal transduction by investigating protein interactions on a spatial grid. More specifically, CA investigates how spatial protein distributions evolve over time, assuming proteins to diffuse along grid points and to react with each other at discrete time steps according to a set of given interaction rules (Bonchev et al. 2010). Siehs and colleagues were the first to introduce the concept of CA to define the complex dynamics of the interplay of pro- and anti-apoptotic Bcl-2 proteins in mitochondrial apoptosis pathway (Siehs et al. 2002). Chen et al. also used this methodology to explore bistability (Chen et al. 2007a). Whereas these studies were solely theoretical, Duessmann et al. (2010) combined the CA approach with experimental data to re-model and quantify Bax translocation and oligomerisation on the mitochondrial membrane in single cells. Furthermore, this study provided an insight into MOMP induction and the rapid pore formation kinetics. The model input was defined as the translocation of the pro-apoptotic protein Bax to the mitochondrial outer membrane, which they determined by quantitative confocal microscopy experiments of yellow fluorescent protein-Bax accumulation in Bax- and Bak-deficient human cells. To understand the kinetics of the reactions leading to MOMP, the outer membrane of a mitochondrion was modelled by a uniformly spaced grid of 100 × 100 squares, with each point on the square having 0–7 proteins positioned at each point, and each protein was given a diffusion probability. The diffusion of proteins between adjacent grid spaces was random, unless the grid was fully occupied in which case diffusion in this direction was blocked. Proteins co-located on the same grid were allowed to react and to produce a set of products according to the known reaction rules of Bcl-2 family members. The likelihood of occurrence of any given specific reaction was estimated, equivalent to the use of kinetic constants in ODE-based models. Stochastic variations were taken into account, generating a detailed description of the dynamics of the reaction over time. The development of pores on the mitochondrial outer membranes occurs on the formation of Bax or Bak tetramers (Schlesinger and Saito 2006) and thus their formation served as the model output. Figure 4.1 illustrates the spatial distribution and temporal evolution of Bax oligomers and Bax bound to pro-survival proteins remodelled from Dussmann et al. (2010). This cellular automaton approach in combination with single cell experimental data has provided important insights on the kinetics of Bax/Bak pore formation at the mitochondrial outer membrane. Of note, it was observed that MOMP was induced in the presence of only minimal levels of Bax and that the majority of Bax translocation and oligomerisation occurred downstream of MOMP. This result has been confirmed by other groups who have shown that pore formation occurs in the presence of nominal levels of BH3-only proteins or Bax (Kuwana et al. 2002; Lovell et al. 2008; Annis et al. 2005). These results indicate that an excess level of Bax may be required to overcome any possible signaling anomalies, such as accumulation of anti-apoptotic Bcl-2 family members at levels sufficient to interfere with pore formation. Alternatively, Bax may have other functions during apoptosis that require higher Bax levels, such as the regulation of mitochondrial fission/fusion.

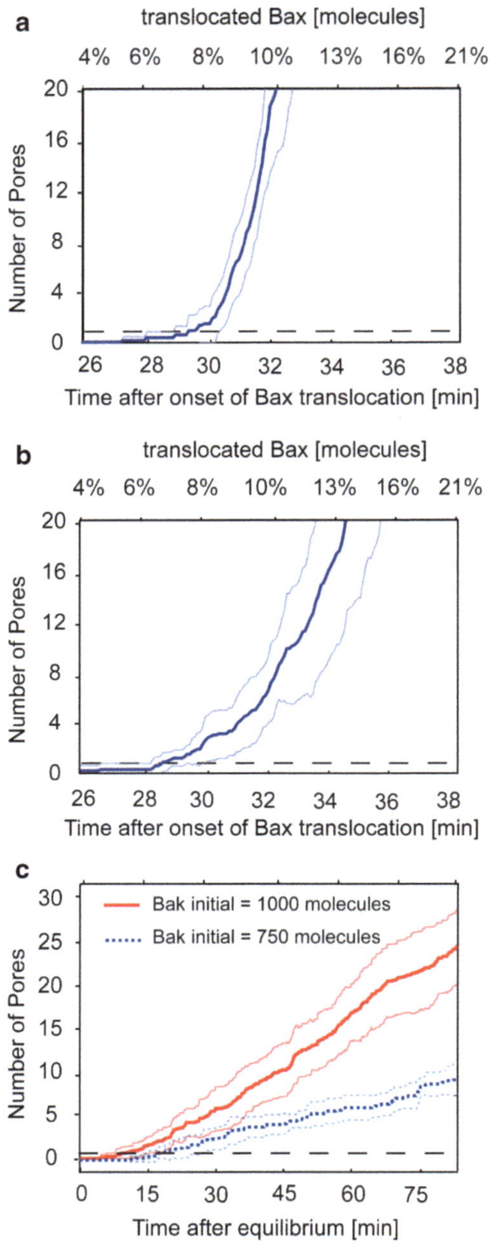

Fig. 4.1 Comparison of direct and indirect MOMP activation models by Cellular Automata (CA): Modelled pore formation for the direct (**a**) and indirect (**b**) activation model with parameters and experimental quantifications taken from [75]. Temporal profiles of pore formation (Bax tetramers) were calculated assuming stochastic movements and reactions. Average (*thick lines*) and intervals for single standard deviation (*thin lines*) were given for 10 calculations. The dashed line indicates pore formation (1 pore). Pore formation with minimal Bax levels was successfully remodelled with the direct and indirect activation model. (**c**) Pore formation in the absence of stimulus in cells expressing 750 (*blue dotted thick* and *thin lines*) or 1,000 Bak molecules (*red solid lines*) per membrane suggests instability of the indirect activation model used in **b**. Stability to pore formation within 24 h was achieved with less than 100 Bak molecules present at the mitochondrial membrane (not shown)

4.5.1 Modeling of the Direct vs. Indirect Pathway of Bax Activation Using CA

Although it is assumed that BH3-only proteins promote apoptosis by binding and inhibiting the activity of anti-apoptotic members of the family, recent analysis identified a possible role as direct enhancers of the pro-apoptotic function of Bax and Bak. The direct activation model activates Bax and Bak directly through interaction with a subclass of BH3-only proteins (PUMA, tBid and Bim) referred to as "activators" (Letai et al. 2002). In comparison, a second subset of BH3-only proteins termed "enablers" function to neutralise anti-apoptotic Bcl-2 proteins through competitive binding, The indirect model, in contrast, assumes that Bax and Bak are bound in a constitutively active state by the anti-apoptotic Bcl-2 family members, and that Bax and Bak are only activated through competitive interactions through their interaction with BH3-only proteins (Annis et al. 2005; Leber et al. 2007).

In fact, a cellular automation approach was used to model both these opposing theories (Chen et al. 2007b; Dussmann et al. 2010). In both models heterodimerisation with the anti-apoptotic Bcl-2 proteins was sufficient to completely inhibit all Bax fractions. In the "direct" model Bax translocates to the mitochondria and consequently is fully activated by activators, resulting in oliogmerisation and pore formation at the mitochondrial membrane. In contrast, in the indirect activation model, Bax was assumed to be activated spontaneously, since no activator is assumed in this model. Thus examination of the Bax pore-forming kinetics of the two competing models informs us that both models can explain for rapid pore formation kinetics under minimal mitochondrial Bax accumulation. Conversely, in order for the indirect model to be an agreement with experimental data, active Bax would be required to be formed by three events per 1,000 Bax molecules per step. However, mathematical modeling suggested that under these conditions, the "indirect" activation model lacks stability, as spontaneous pore formation and apoptosis are predicted to occur already in non-stimulated cells (Fig. 4.1) (Dussmann et al. 2010).

4.6 Conclusions

Mathematical modeling of biological process is a developing method used to provide novel insights into the complex pathways, such as the apoptosis pathway. These models have the potential to predict and explain cellular behaviour in a quantitative manner which cannot be achieved by biological data alone. To date, elucidating the complex interplay between pro- and anti-survival Bcl-2 proteins using a modeling approach has been limited due to the lack of comprehensive biochemical data defining their structural activation within biological membranes. It is anticipated that maturation of the modeling approaches can elucidate these mechanisms and enhance the development and specificity of therapeutics such as Bcl-2 antagonists. Experimentally, defining the transcriptional and post-translational activation of BH3-only proteins at the single cell level will be of particular significance to the generation of models of stress and survival signaling pathways as these activation mechanisms could serve as input, intermediary or output of a system. Finally, combining experimental measurements with mathematical modeling has the potential to provide mechanistic insights into the individual factors underlying disease progression. The translation of mathematical models to a clinical setting for predicting responsiveness to therapies may enable tailored treatments for individual patients.

Acknowledgements We thank Dr. Heiko Duessmann and Dr. Jakub Wenus for data shown in Fig. 4.1. This research was supported by the Health Research Board grant, TRA/2007/26; the National Biophotonics and Imaging Platform (HEA PRTLI Cycle 4) and the EU Framework Programme 7 (Project APO-SYS).

References

Acehan D et al (2002) Three-dimensional structure of the apoptosome: implications for assembly, procaspase-9 binding, and activation. Mol Cell 9(2):423–432

Adams JM, Cory S (2007) Bcl-2-regulated apoptosis: mechanism and therapeutic potential. Curr Opin Immunol 19(5):488–496

Adrain C et al (1999) Regulation of apoptotic protease activating factor-1 oligomerization and apoptosis by the WD-40 repeat region. J Biol Chem 274(30):20855–20860

Albeck JG et al (2008a) Quantitative analysis of pathways controlling extrinsic apoptosis in single cells. Mol Cell 30(1):11–25

Albeck JG et al (2008b) Modeling a snap-action, variable-delay switch controlling extrinsic cell death. PLoS Biol 6(12):2831–2852

Annis MG et al (2005) Bax forms multispanning monomers that oligomerize to permeabilize membranes during apoptosis. EMBO J 24(12):2096–2103

Bentele M et al (2004) Mathematical modelling reveals threshold mechanism in CD95-induced apoptosis. J Cell Biol 166(6):839–851

Boatright KM et al (2003) A unified model for apical caspase activation. Mol Cell 11(2):529–541

Bonchev D et al (2010) Cellular automata modelling of biomolecular networks dynamics. SAR QSAR Environ Res 21(1):77–102

Brown JM, Wouters BG (1999) Apoptosis, p53, and tumor cell sensitivity to anticancer agents. Cancer Res 59(7):1391–1399

Chai J et al (2001) Structural basis of caspase-7 inhibition by XIAP. Cell 104(5):769–780

Chen L et al (2005) Differential targeting of prosurvival Bcl-2 proteins by their BH3-only ligands allows complementary apoptotic function. Mol Cell 17(3):393–403

Chen C et al (2007a) Modeling of the role of a Bax-activation switch in the mitochondrial apoptosis decision. Biophys J 92(12):4304–4315

Chen C et al (2007b) Robustness analysis identifies the plausible model of the Bcl-2 apoptotic switch. FEBS Lett 581(26):5143–5150

Danial NN, Korsmeyer SJ (2004) Cell death: critical control points. Cell 116(2):205–219

Deveraux QL, Reed JC (1999) IAP family proteins–suppressors of apoptosis. Genes Dev 13(3):239–252

Deveraux QL et al (1997) X-linked IAP is a direct inhibitor of cell-death proteases. Nature 388(6639):300–304

Deveraux QL et al (1998) IAPs block apoptotic events induced by caspase-8 and cytochrome c by direct inhibition of distinct caspases. EMBO J 17(8):2215–2223
Donepudi M, Grutter MG (2002) Structure and zymogen activation of caspases. Biophys Chem 101–102:145–153
Dussmann H et al (2010) Single-cell quantification of Bax activation and mathematical modelling suggest pore formation on minimal mitochondrial Bax accumulation. Cell Death Differ 17(2):278–90
Eissing T et al (2004) Bistability analyses of a caspase activation model for receptor-induced apoptosis. J Biol Chem 279(35):36892–36897
Fussenegger M, Bailey JE, Varner J (2000) A mathematical model of caspase function in apoptosis. Nat Biotechnol 18(7):768–774
Goldstein JC et al (2000) The coordinate release of cytochrome c during apoptosis is rapid, complete and kinetically invariant. Nat Cell Biol 2(3):156–162
Hengartner MO, Horvitz HR (1994) Activation of C elegans cell death protein CED-9 by an amino-acid substitution in a domain conserved in Bcl-2. Nature 369(6478):318–320
Hockenbery D et al (1990) Bcl-2 is an inner mitochondrial membrane protein that blocks programmed cell death. Nature 348(6299):334–336
Hofer-Warbinek R et al (2000) Activation of NF-κ B by XIAP, the X chromosome-linked inhibitor of apoptosis, in endothelial cells involves TAK1. J Biol Chem 275(29):22064–22068
Huang H et al (2000) The inhibitor of apoptosis, cIAP2, functions as a ubiquitin-protein ligase and promotes in vitro monoubiquitination of caspases 3 and 7. J Biol Chem 275(35):26661–26664
Huber HJ et al (2009) TOXI-SIM-A simulation tool for the analysis of mitochondrial and plasma membrane potentials. J Neurosci Methods 176(2):270–275
Huber HJ et al (2011) Glucose metabolism determines resistance of cancer cells to bioenergetic crisis after cytochrome-c release. Mol Syst Biol 7:470
Kaufman RJ (1999) Stress signaling from the lumen of the endoplasmic reticulum: coordination of gene transcriptional and translational controls. Genes Dev 13(10):1211–1233
Kelley SK, Ashkenazi A (2004) Targeting death receptors in cancer with Apo2L/TRAIL. Curr Opin Pharmacol 4(4):333–339
Kluck RM et al (1997a) The release of cytochrome c from mitochondria: a primary site for Bcl-2 regulation of apoptosis. Science 275(5303):1132–1136
Kluck RM et al (1997b) Cytochrome c activation of CPP32-like proteolysis plays a critical role in a Xenopus cell-free apoptosis system. EMBO J 16(15):4639–4649
Kohler B et al (2008) Bid participates in genotoxic drug-induced apoptosis of HeLa cells and is essential for death receptor ligands' apoptotic and synergistic effects. PLoS One 3(7):e2844
Kuwana T et al (2002) Bid, Bax, and lipids cooperate to form supramolecular openings in the outer mitochondrial membrane. Cell 111(3):331–342
Lavrik IN et al (2007) Analysis of CD95 threshold signaling: triggering of CD95 (FAS/APO-1) at low concentrations primarily results in survival signaling. J Biol Chem 282(18):13664–13671
Leber B, Lin J, Andrews DW (2007) Embedded together: the life and death consequences of interaction of the Bcl-2 family with membranes. Apoptosis 12(5):897–911
Letai A et al (2002) Distinct BH3 domains either sensitize or activate mitochondrial apoptosis, serving as prototype cancer therapeutics. Cancer Cell 2(3):183–192
Li P et al (1997) Cytochrome c and dATP-dependent formation of Apaf-1/caspase-9 complex initiates an apoptotic protease cascade. Cell 91(4):479–489
Li H et al (1998) Cleavage of BID by caspase 8 mediates the mitochondrial damage in the Fas pathway of apoptosis. Cell 94(4):491–501
Liston P, Fong WG, Korneluk RG (2003) The inhibitors of apoptosis: there is more to life than Bcl2. Oncogene 22(53):8568–8580
Liu X et al (1996) Induction of apoptotic program in cell-free extracts: requirement for dATP and cytochrome c. Cell 86(1):147–157
Lovell JF et al (2008) Membrane binding by tBid initiates an ordered series of events culminating in membrane permeabilization by Bax. Cell 135(6):1074–1084

Lu M et al (2007) XIAP induces NF-κB activation via the BIR1/TAB1 interaction and BIR1 dimerization. Mol Cell 26(5):689–702
Luo X et al (1998) Bid, a Bcl2 interacting protein, mediates cytochrome c release from mitochondria in response to activation of cell surface death receptors. Cell 94(4):481–490
MacFarlane M et al (2002) Proteasome-mediated degradation of Smac during apoptosis: XIAP promotes Smac ubiquitination in vitro. J Biol Chem 277(39):36611–36616
Martins LM et al (2002) The serine protease Omi/HtrA2 regulates apoptosis by binding XIAP through a reaper-like motif. J Biol Chem 277(1):439–444
Neumann L et al (2010) Dynamics within the CD95 death-inducing signaling complex decide life and death of cells. Mol Syst Biol 6:352
O'Connor CL et al (2008) Intracellular signaling dynamics during apoptosis execution in the presence or absence of X-linked-inhibitor-of-apoptosis-protein. Biochim Biophys Acta 1783 (10):1903–1913
Oltvai ZN, Milliman CL, Korsmeyer SJ (1993) Bcl-2 heterodimerizes in vivo with a conserved homolog, Bax, that accelerates programmed cell death. Cell 74(4):609–619
Pace V et al (2010) Experimental testing of a mathematical model relevant to the extrinsic pathway of apoptosis. Cell Stress Chaperones 15(1):13–23
Rehm M, Dussmann H, Prehn JH (2003) Real-time single cell analysis of Smac/DIABLO release during apoptosis. J Cell Biol 162(6):1031–1043
Rehm M et al (2006) Systems analysis of effector caspase activation and its control by X-linked inhibitor of apoptosis protein. EMBO J 25(18):4338–4349
Ricci JE, Gottlieb RA, Green DR (2003) Caspase-mediated loss of mitochondrial function and generation of reactive oxygen species during apoptosis. J Cell Biol 160(1):65–75
Riedl SJ et al (2001) Structural basis for the inhibition of caspase-3 by XIAP. Cell 104(5):791–800
Rodriguez J, Lazebnik Y (1999) Caspase-9 and APAF-1 form an active holoenzyme. Genes Dev 13(24):3179–3184
Salvesen GS, Duckett CS (2002) IAP proteins: blocking the road to death's door. Nat Rev Mol Cell Biol 3(6):401–410
Sanna MG et al (1998) Selective activation of JNK1 is necessary for the anti-apoptotic activity of hILP. Proc Natl Acad Sci USA 95(11):6015–6020
Schlesinger PH, Saito M (2006) The Bax pore in liposomes, Biophysics. Cell Death Differ 13(8):1403–1408
Shi Y (2006) Mechanical aspects of apoptosome assembly. Curr Opin Cell Biol 18(6):677–684
Shiozaki EN et al (2003) Mechanism of XIAP-mediated inhibition of caspase-9. Mol Cell 11(2):519–527
Siehs C et al (2002) Discrete simulation of regulatory homo- and heterodimerization in the apoptosis effector phase. Bioinformatics 18(1):67–76
Slee EA et al (1999) Ordering the cytochrome c-initiated caspase cascade: hierarchical activation of caspases-2, -3, -6, -7, -8, and −10 in a caspase-9-dependent manner. J Cell Biol 144(2):281–292
Srinivasula SM et al (2001) A conserved XIAP-interaction motif in caspase-9 and Smac/DIABLO regulates caspase activity and apoptosis. Nature 410(6824):112–116
Stucki JW, Simon HU (2005) Mathematical modeling of the regulation of caspase-3 activation and degradation. J Theor Biol 234(1):123–131
Varfolomeev E et al (2007) IAP antagonists induce autoubiquitination of c-IAPs, NF-κB activation, and TNFalpha-dependent apoptosis. Cell 131(4):669–681
Verhagen AM et al (2000) Identification of DIABLO, a mammalian protein that promotes apoptosis by binding to and antagonizing IAP proteins. Cell 102(1):43–53
Vucic D, Kaiser WJ, Miller LK (1998) Inhibitor of apoptosis proteins physically interact with and block apoptosis induced by Drosophila proteins HID and GRIM. Mol Cell Biol 18(6):3300–3309
Wei MC et al (2001) Proapoptotic BAX and BAK: a requisite gateway to mitochondrial dysfunction and death. Science 292(5517):727–730

Yang Y et al (2000) Ubiquitin protein ligase activity of IAPs and their degradation in proteasomes in response to apoptotic stimuli. Science 288(5467):874–877

Youle RJ, Strasser A (2008) The BCL-2 protein family: opposing activities that mediate cell death. Nat Rev Mol Cell Biol 9(1):47–59

Zhang T, Brazhnik P, Tyson JJ (2009) Computational analysis of dynamical responses to the intrinsic pathway of programmed cell death. Biophys J 97(2):415–434

Zou H et al (1997) Apaf-1, a human protein homologous to C elegans CED-4, participates in cytochrome c-dependent activation of caspase-3. Cell 90(3):405–413

Zou H et al (2003) Regulation of the Apaf-1/caspase-9 apoptosome by caspase-3 and XIAP. J Biol Chem 278(10):8091–8098

Chapter 5
Systems Biology of Cell Death in Hepatocytes

Rebekka Schlatter*, Kathrin Schmich*, Christoph Borner, Michael Ederer, and Irmgard Merfort

Abstract The processes of liver regeneration and liver disease are regulated by a complex network of soluble and cell-associated apoptotic and inflammatory signals. To obtain insights into the mechanistic interplay of these signals and to define new therapeutic strategies, the combination of experimental data and mathematical modeling is a promising systems biological approach. Here, we review recent results in death receptor-mediated hepatocyte apoptosis focusing on Fas/CD95 and TNFα-mediated cell death. In this context, we present two complementary approaches of modeling death receptor-mediated cell death in hepatocytes. On the one hand we describe an ODE model of TNFα and FasL sensitising, which was extended by adding the regulation of pJNK and the generation of ROS after combined TNFα and ActD treatment and in which a published NF-κB model was integrated. This model is suitable for the integration of further pathway models, thus contributing to a better understanding of the network. On the other hand a literature-based and in parts experimentally validated comprehensive

*Rebekka Schlatter and Kathrin Schmich contributed equally.

R. Schlatter • M. Ederer
Institute for System Dynamics, University of Stuttgart, Pfaffenwaldring 9, 70569 Stuttgart, Germany

K. Schmich • I. Merfort (✉)
Department of Pharmaceutical Biology and Biotechnology, Albert Ludwigs University Freiburg, Stefan Meier Str. 19, 79104 Freiburg, Germany
e-mail: irmgard.merfort@pharmazie.uni-freiburg.de

C. Borner
Institute of Molecular Medicine and Cell Research, Albert Ludwigs University Freiburg, 79104 Freiburg, Germany

Spemann Graduate School of Biology and Medicine (SGBM), Albert Ludwigs University Freiburg, Albertstrasse 19a, 79106 Freiburg, Germany

Bioss—Centre for Biological Signaling Studies, Albert Ludwigs University Freiburg, Albertstrasse 19, 79106 Freiburg, Germany

I.N. Lavrik (ed.), *Systems Biology of Apoptosis*, DOI 10.1007/978-1-4614-4009-3_5,

Boolean model of the central intrinsic and extrinsic apoptosis pathways as well as pathways connected with them is described. In the future, the according mathematical models will be a valuable approach to understand the complex crosstalks and interactions within hepatocytes and between different cells in the liver. Thus, modeling of apoptosis in hepatocytes will proceed on different routes on the way to a functional representation of the whole liver.

5.1 Introduction

The liver is an organ of immense complexity and essential for survival since no other organ can compensate its multiplicity of functions. Among these, the synthesis, metabolism, storage and redistribution of fats, carbohydrates, vitamins and nutrients are of special relevance. Furthermore, the liver is the main detoxifying organ of the body. Eighty percent of the liver mass comprises the parenchymal cell, the hepatocyte, while the remaining 20% are non-parenchymal cells such as the innate immune system-related Kupffer cells, which are liver-located macrophages, or lymphocytes, but also stellate cells or endothelial cells (Taub 2004). The liver has an amazing regenerative capacity which allows recovering lost liver mass up to 70% of its original volume. Normally hepatocytes, bile duct epithelium and endothelial cells are involved in liver regeneration, whereas stem cells do not play a role (Papa et al. 2009).

The process of liver regeneration is complex and characterised by several strongly interacting pathways (Taub 2004; Fausto et al. 2006; Michalopoulos 2007; Mohammed and Khokha 2005). Disruptions in the regulatory mechanisms can totally impair the whole process and result in liver injury and liver failure. It is generally accepted that enhanced apoptosis induced by death receptor ligands such as TNFα and FasL is critically implicated in the mechanisms and is linked to a wide range of liver diseases. Dysregulation can also lead to excessive proliferation and finally to hepatocellular carcinoma (HCC). Most often, hepatocytes are involved in liver injury including viral hepatitis, alcoholic liver diseases, acute liver failure, ischaemia-reperfusion injury, graft rejection and diseases of the bile ducts (Papa et al. 2009; Hatano 2007; Malhi and Gores 2008; Akazawa and Gores 2007; Malhi et al. 2010).

The processes of liver regeneration and liver disease are regulated by a complex network of soluble and cell-associated apoptotic and inflammatory signals (Malhi and Gores 2008). It is therefore increasingly important to obtain insight into the mechanistic interplay of these signals to define new therapeutic strategies. Mathematical modeling is a helpful tool to understand and analyse the dynamical behaviour of such a complex regulatory network (Calzone et al. 2010). Considering the pivotal importance of the liver in human health and disease, the German ministry for education and research has started in 2004 the financial support of the project "HepatoSys" which focused on the investigation of the liver in a systems biology approach. In the beginning the project concentrated on cell signaling pathways in primary hepatocytes. The follow-up project "Virtual Liver" has now been extended to the tissue level and aims to understand and model a complete organ from the biochemical, cellular, to the whole organ level.

In the course of the above-mentioned projects, new insights have been gained in primary hepatocytes on signaling pathways which are involved in the decision of survival and apoptosis by a systems biological approach (Philippi et al. 2009; Schlatter et al. 2009a, 2011; Schmich et al. 2011). As a model, primary murine hepatocytes were chosen, which are more difficult to be handled as compared to standard cell lines. A standard operating procedure (SOP) for the isolation and cultivation of primary murine hepatocytes has been established, which is a prerequisite for the development of mathematical models (Klingmuller et al. 2006). Here, we review recent results in death receptor-mediated hepatocyte apoptosis focusing on Fas/CD95- and TNFα-mediated cell death. At first the biological background and then the modeling approach are presented.

5.2 Apoptosis Pathways in the Liver Focusing on Hepatocytes

5.2.1 Fas Signaling in the Liver

FasL/CD95L and its corresponding receptor Fas/CD95 play a pivotal role in the immune system inducing the death of virally infected cells as well as of obsolete lymphocytes (Li-Weber and Krammer 2003). Defects in this system can favour the development of autoimmune diseases. In the human body, the highest constitutive expression of Fas is found on the surface of hepatocytes where it is critically involved in hepatic health and diseases (Peter et al. 2007). Hence, mice treated with a lethal dose of agonistic anti-Fas antibody die due to massive hepatocyte apoptosis and liver failure (Ogasawara et al. 1993). Moreover, liver cells are thought to die through Fas-mediated apoptosis during viral and autoimmune hepatitis, alcoholic liver disease and endotoxin- or ischaemia/reperfusion-induced liver damage (Malhi and Gores 2008; Malhi et al. 2010; Canbay et al. 2004).

The molecular mechanism of Fas-induced apoptosis has widely been studied in several cell types. Engagement of Fas by FasL results in receptor aggregation and trimerization (Lavrik et al. 2007). In the next step, the adaptor molecule FADD associates with the death domain (DD) of Fas and recruits procaspase-8. This complex of proteins is called the death-inducing signaling complex (DISC) (Hughes et al. 2009). At this point, two different Fas signaling pathways can be executed (Scaffidi et al. 1998), the so-called type I and type II pathways. How the decision between the two pathways is undertaken remains currently unclear. There are suggestions that the amount of activated caspase-8 decides for the substrate specificity (Krammer 2000), while others propose differences in Fas aggregation and internalisation, the amount of substrates that must be proteolysed or the levels of caspase inhibitors such as XIAP to be critical (Strasser et al. 2009).

In general, Fas apoptosis signaling can be classified to be Bid independent (type I signaling) or Bid dependent (type II pathway) in several cell types. Gene-targeting experiments have shown that caspase-8-mediated activation of Bid is essential in pancreatic β-cells, while it is dispensable in lymphoid cells (McKenzie et al. 2008; Kaufmann et al. 2007). In hepatocytes, Fas-induced cell death is dependent on Bid since Bid-deficient mice are resistant to Fas-induced hepatocellular apoptosis, fulminant hepatitis and subsequent liver failure (Yin et al. 1999). Thus, in vivo Fas-mediated apoptosis in hepatocytes occurs via the Bid-dependent type II pathway. However, recent studies have weakened the fixed definitions of type I and type II pathway by demonstrating that Fas-induced type II apoptosis can switch to type I upon different culturing conditions in primary hepatocytes (Walter et al. 2008). Furthermore, it was shown that high concentrations of FasL may also result in type I signaling in vivo (Schungel et al. 2009).

Remarkably, Fas is not only a potent inducer of apoptosis but can also activate non-apoptotic pathways. In this context, Fas turned out to be critical for liver regeneration after partial hepatectomy (PH). The mechanism remains unclear, but nuclear factor kappaB (NF-κB) and MAP kinase survival signaling have been implicated (Peter et al. 2007). Furthermore, hepatocyte-specific loss of caspase-8 also results in impaired liver regeneration after PH (Ben et al. 2007). The role of caspase-8 in this process again is complex. Caspase-8 has been shown to induce, together with FADD and RIP, potent NF-κB activation upon Fas treatment, while cFLIP inhibits this activation (Kreuz et al. 2004). In addition, procaspase-8 generates the cFLIP cleavage products p43-FLIP and p22-FLIP which both also activate NF-κB (Neumann et al. 2010). Hence, mediators that have originally only been associated with apoptosis induction turn out to be relevant for non-apoptotic functions, and the physiological role of Fas extends far beyond cell death control (Peter et al. 2007).

5.2.2 TNFα-Mediated Apoptosis in Hepatocytes

TNFα is a pleiotropic cytokine that induces a variety of cellular responses such as inflammation and cell proliferation mainly by activation of NF-κB signaling cascade. Unlike FasL, association of TNFα with its main receptor TNFR1 does not primarily lead to cell death in most cell types, including hepatocytes (Varfolomeev and Ashkenazi 2004). Nevertheless, TNFα has been assigned a role in the carcinogenetic process (Karin 2006) as well as in chronic viral hepatitis. Chronic hepatitis C infection is known to cause persistent inflammation and fibrosis in the liver which is characterised by increased TNFα levels and ongoing NF-κB activation (Papa et al. 2009; Malhi et al. 2010; Sun and Karin 2008). This chronic inflammatory stage often results in liver cirrhosis or liver cancer due to protection from apoptosis as well as NF-κB-induced enhanced proliferation. TNFα can additionally be a main mediator of hepatocyte apoptosis and liver failure (Bradham et al. 1998). This is the case in alcoholic liver disease. Elevated serum levels of TNFα have been correlated with increased lethality, and TNFR1 has been shown to be essential for alcohol-induced liver injury (Malhi and Gores 2008). The role of TNFα in liver apoptosis has widely been studied in the LPS/GalN model. D-Galactosamine (GalN) is an amino-sugar which is, via its metabolism in the liver, able to selectively inhibit hepatic transcription. Endotoxins such as lipopolysaccharide (LPS) are used to initiate the inflammatory response in this model. Importantly, it has been shown that LPS/GalN-mediated liver injury depends on TNFα and its receptor TNFR1 (Nowak et al. 2000). Accordingly, this model is often used to study TNFα-mediated liver apoptosis in vivo.

Various studies have been undertaken to define the role of TNFα signaling in hepatocyte apoptosis [for literatures, see review (Papa et al. 2009; Malhi et al. 2010)]. Upon binding of TNFα to TNFR1, a membrane bound complex I comprising of TNFR1, receptor-interacting protein 1 (RIP1), TNF-receptor-associated factor 2 (TRAF2) and TNFR1-associated death domain protein (TRADD) is first formed which rapidly activates the survival transcription factor NF-κB and the c-Jun *N*-terminal kinase (JNK) (Micheau and Tschopp 2003; Papa et al. 2006). To signal for cell death, a second, receptor-free complex II has to assemble in the cytoplasm which still contains RIP1, TRAF2 and TRADD but recruits FADD and procaspase-8 to activate procaspase-3/-7 (Micheau and Tschopp 2003). Under normal conditions, complex II formation is blocked by cFLIP and NF-κB survival signaling (Karin and Lin 2002). TNFα-mediated NF-κB activation relies on the IκB kinase complex (IKK). IKK becomes activated by assembly with TRAF2 and RIP1 and subsequently phosphorylates the inhibitor of κB (IκB) proteins which normally restrain NF-κB within the cytosol (Wajant et al. 2003). The phosphorylated IκB proteins become ubiquitinylated and degraded by the proteasome, a process that allows free NF-κB to translocate to the nucleus and initiate transcription of target genes (Baud and Karin 2001). The transcription of anti-apoptotic genes is supposed to be the main reason why TNFα does not induce cell death in most cell types. Among these genes are the direct inhibitors of apoptosis Bcl-x_L, XIAP, cIAP1 and 2 and cFLIP (Baud and Karin 2009; Micheau et al. 2001). cIAP1 and 2 belong to the protein family of inhibitor of apoptosis proteins (IAPs) and are associated to TRAF2 at the TNF receptor. Although their direct function in cell survival is unknown, neutralisation of cIAPs by the respective antagonists results in TNFα-triggered apoptosis (Vince et al. 2007).

However, the factors mentioned above cannot fully account for the anti-apoptotic activity of NF-κB. Another important mechanism to protect cells from TNFα-induced apoptosis is the inhibition of pro-apoptotic JNK signaling. JNK is a serine/threonine protein MAPK which is known to be rapidly activated by TNF complex I via a MAPK kinase cascade (Lin 2003). This transient activation is rapidly terminated by MAPK phosphatases (MKPs) which are themselves controlled by NFκB (Papa et al. 2006). However, under certain conditions, TNFα can lead to a sustained activation of JNK which causes apoptosis. In this case, the termination mechanism of the JNK activity is disturbed, as the activity of MKPs is inhibited by reactive oxygen species (ROS) (Papa et al. 2006; Kamata et al. 2005;

Nakano et al. 2006). The origin of ROS in response to TNFR activation is still unknown, but it is assumed that ROS are generated primarily in the mitochondria during the transport of electrons in the respiratory chain (Nakano et al. 2006). However a recent study has also implicated the plasma membrane bound NADPH oxidase as an early producer of ROS after TNFα treatment (Yazdanpanah et al. 2009). TNFα-induced ROS can accumulate under the condition that they are no longer suppressed by antioxidant enzymes, such as superoxide dismutases (SODs), glutathione peroxidase or ferritin. When ROS accumulation induces prolonged activity of JNK, the connection of this pathway to cell death execution has been proposed. On the one hand, JNK can directly activate the E3 ubiquitin ligase Itch which induces proteasomal processing of cFLIP (Chang et al. 2006). Reduced cFLIP levels allow procaspase-8 to assemble with complex I and form the pro-apoptotic complex II. Interestingly, this mechanism only works in the presence of TNFα and the translation inhibitor cycloheximide. No cFLIP degradation is observed combining TNFα with the transcription inhibitor actinomycin D (ActD) (Schlatter et al. 2011). Furthermore, it has been proposed that JNK can directly activate the mitochondrial pathway. While one study suggested that JNK generates the pro-apoptotic Bid cleavage product jBid (Deng et al. 2003), several other reports indicate that JNK activates Bim by phosphorylation (Papa et al. 2009; Corazza et al. 2006; Kaufmann et al. 2009). Both Bid and Bim can trigger mitochondrial apoptosis type II pathway shown among others in a model of fatal hepatitis. Thus, apoptosis induction by TNFα is tightly regulated by a complex network of signaling pathways and dysfunction of any of these signals affects the whole system.

5.2.3 FasL/TNFα Signaling in Hepatocytes

It has been shown that hepatic Fas and circulating FasL levels are enhanced in alcoholic steatohepatitis and the crosstalk of FasL and TNFα seems to contribute to disease progression (Malhi and Gores 2008). In accordance with these in vivo results, we have recently shown that TNFα sensitises primary murine hepatocytes cultured on collagen to FasL-induced apoptosis (Schmich et al. 2011). The synergism was shown to be time dependent and specifically mediated by TNFα. Fas itself was essential for the sensitization, but neither Fas upregulation nor endogenous FasL induction was responsible for this effect. Whereas FasL was confirmed to induce Bid-independent apoptosis in hepatocytes cultured on collagen, the sensitising effect of TNFα was clearly dependent on Bid. Furthermore, both JNK activation and the BH3-only protein Bim play a crucial role in TNFα-induced apoptosis sensitization. The interplay of Bim and Bid activates the mitochondrial amplification loop and induces cytochrome c release, a hallmark of type II apoptosis. Thus, our data confirmed that TNFα is capable of engaging the JNK/Bim apoptotic pathway and of restoring type II signaling on collagen-cultured primary

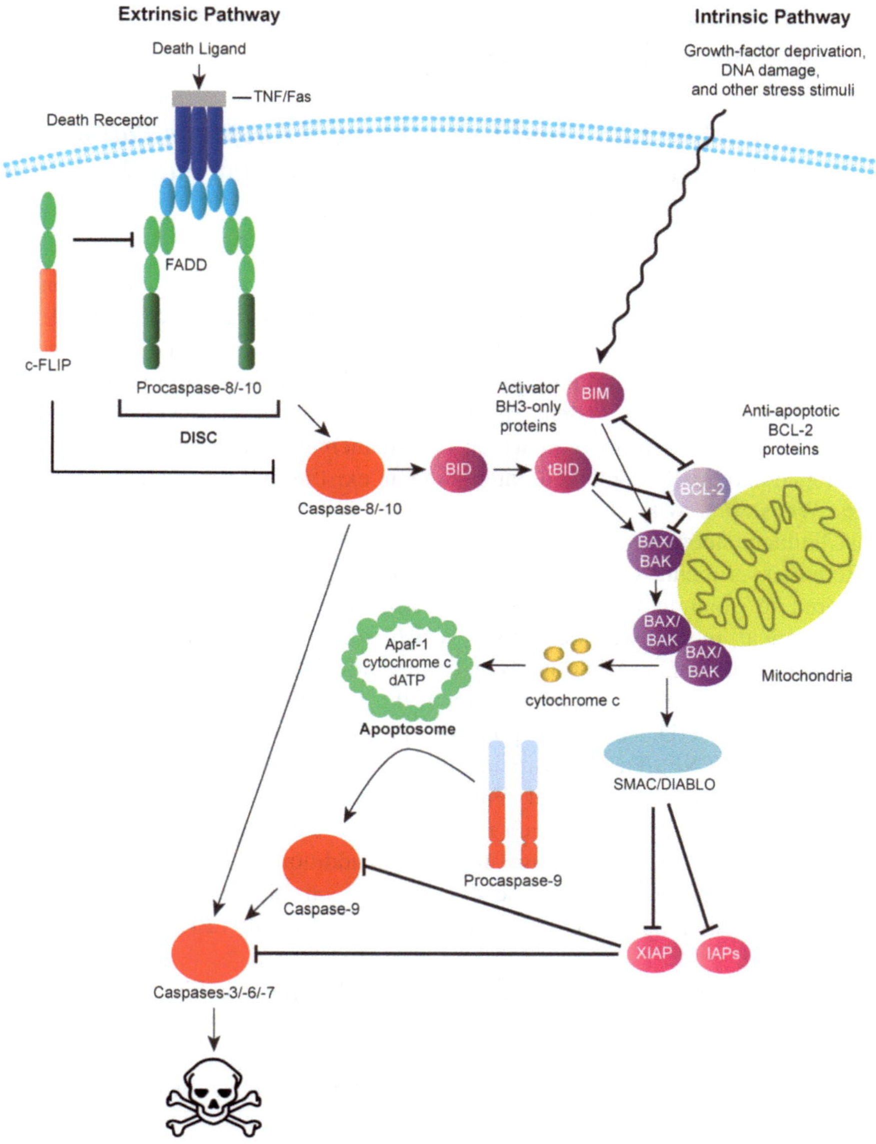

Fig. 5.1 Intrinsic and extrinsic apoptotic pathways according to Plati et al. (2011)

hepatocytes. This sensitising effect towards FasL-induced liver damage was also demonstrated in vivo. Interestingly, it was even reported that loss of TNFR1 and TNFR2 protects mice from anti-Fas antibody-induced liver injury (Costelli et al. 2003) (Fig. 5.1).

5.3 Modeling Approaches of Apoptosis Pathways in Primary Hepatocytes

5.3.1 Modeling FasL- and TNFα-Induced Apoptosis Pathways Using ODE Models as an Example for Studying Qualitative Mechanisms in a Complex Interplay

As mentioned above detailed knowledge on individual molecule interactions in apoptosis signaling is available. This is a prerequisite for mathematical modeling based on kinetic rate equations and differential equations and a powerful tool to combine the knowledge on the molecule interactions with the aim to explain the overall behaviour of the system. In this subchapter we provide an example for a successful modeling process in which two mathematical apoptosis models were developed and finally combined, integrating an existing NF-κB model from literature. We started with the description of the TNFα sensitising effect towards FasL-induced apoptosis. This effect was first analysed in detail in the context of different experimental conditions and genotypes. Sufficient measurement data were gathered to point to a certain crosstalk structure. Based on this hypothesis a mathematical model was developed which successfully reproduced the data (Schmich et al. 2011). Additionally, a mathematical model was developed which reproduced the complex interplay when TNFα was combined with the transcriptional inhibitor ActD, leading to apoptosis in hepatocytes. The model developed to reproduce the TNFα/FasL sensitising was adapted, enhanced substantially and combined with a NF-κB model. Finally, a large ordinary differential equation (ODE) model was obtained which reproduced the data on both crosstalk effects and might serve as core structure for further additions (Schlatter et al. 2011). The full description of the models, technical and biological details as well as according references can be found in Schlatter et al. (2011) and Schmich et al. (2011).

A scheme of the mathematical sensitising model is shown in Fig. 5.2. The model is based on literature and own results, and the model parameters have been fitted to qualitatively reproduce the available experimental data (Schmich et al. 2011).

Hepatocytes treated with Fas ligand only exhibit apoptosis which is strongly increased in XIAP knockout cells as compared to wild-type cells. This is reproduced by the mathematical model via the direct caspase-3 cleavage by activated caspase-8 (v14). The amount of activated caspase-3 is however limited by the protein XIAP which functions as a buffer in wild-type cells (v18). Accordingly, by omitting XIAP in the mathematical model, and thereby mimicking XIAP knockout cells, the caspase-3 activity in the simulation is considerably increased. Moreover, it was shown that in FasL-treated, cultured hepatocytes apoptosis is independent of Bid and occurs without mitochondrial release of cytochrome c (Walter et al. 2008). In the sensitising model, a few Bid molecules are cleaved following FasL treatment (v9). However, this process has no further impact as the cleaved form tBid is

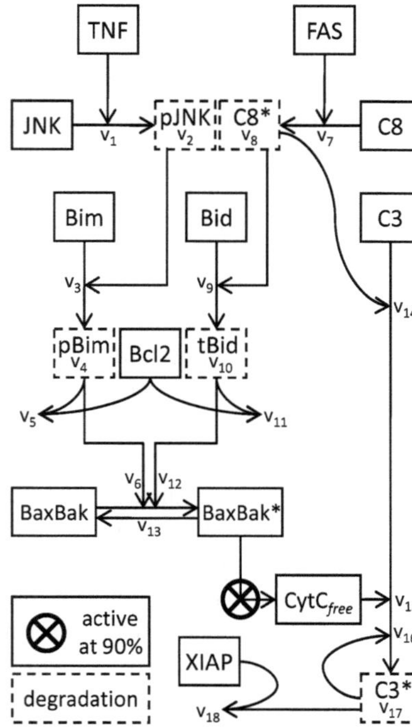

Fig. 5.2 Scheme of the sensitising model, published in Schmich et al. (2011)

buffered by anti-apoptotic members of the Bcl2 family (v11). Thus, only a very low number of Bax and Bak molecules are activated by tBid (v12), which is far below the threshold of 90% Bax/Bak activation needed in the model to result in cytochrome c release.

Cultured hepatocytes treated with TNFα only fail to induce apoptosis, although other cell types are known to die after TNFα treatment. This is realised in the sensitising model where TNFα leads to JNK phosphorylation (v1) and subsequent phosphorylation of Bim (v3). However, as discussed for tBid, also phosphorylated Bim is buffered by members of the Bcl-2 family (v5) and is therefore not sufficient to cause cytochrome c release via Bax/Bak activation (v6). Notably, the presented mechanism allows apoptosis after TNFα treatment in the mathematical model if the initial amounts of the participating species, especially Bim and Bcl-2, are changed. This change in the model corresponds to different expression levels of these proteins in the cell and thereby the presented sensitising model is not contradicting the apoptotic effect of TNFα in other cell types.

The sensitising effect of TNFα towards Fas ligand-induced apoptosis is realised in the mathematical model by the interplay of both signaling pathways on the level of the Bcl-2 buffer. TNFα clears the way for tBid by neutralising the Bcl-2 proteins via phosphorylated Bim (v5). The rate constants of the according interactions were fitted to reproduce the experimentally determined lead time needed for TNFα to achieve the sensitising effect. Beyond neutralising Bcl-2, phosphorylated Bim also directly activates a small amount of Bax/Bak. With the Bcl-2 buffer being consumed, tBid can then activate a sufficient amount of Bax/Bak (v12) to induce cytochrome c release after TNFα and FasL treatment. Importantly, sensitization does not work in the model when FasL is applied first as it was also experimentally shown in hepatocytes. The reason is the temporal progression of Bax/Bak activation which is counteracted by a backwards reaction in the model. If FasL is applied first, tBid consumes the Bcl-2 buffer (v11) and a small amount of Bax/Bak is activated (v12) which is however already decayed before the more slowly signaling after TNFα is able to phosphorylate Bim. In this scenario, the achieved amount of activated Bax/Bak during the considered time span is under the threshold for cytochrome c release.

Using knockout mutants, the TNFα/FasL sensitising effect was shown to be Bid dependent and XIAP independent, but FasL-induced apoptosis alone could not be further increased in $XIAP^{-/-}$ hepatocytes (Schmich et al. 2011). These results are also reproduced by the sensitising model. If Bid is missing in the model, Bax/Bak is not activated to a sufficient amount to result in cytochrome c release. If XIAP is missing in the model, caspase-3 is activated after FasL treatment almost maximally, i.e. to 100% of the initial amount of inactive C3, and therefore cannot be further increased by cytochrome c. Overall, the mathematical sensitising model is able to reproduce our experimental data as well as literature knowledge. This shows that the modelled interactions are sufficient to explain the observed behaviour.

A second mathematical model describing another apoptotic crosstalk in cultured hepatocytes was developed and both models were combined. Although it has been known for a long time that only the combination of ActD and TNFα, but not TNFα alone induces apoptosis in primary hepatocytes, no mathematical model has been established. A scheme of the finally developed enhanced apoptosis model is shown in Fig. 5.3. Here, the core structure of the sensitising model is sustained on the left side of the figure and the overall model still reproduces the sensitising effect as discussed above. However, modeling of TNFα signaling was substantially enhanced and also an existing NF-κB model from Lipniacki et al. was integrated (Lipniacki et al. 2004), which is represented by a grey box in Fig. 5.3.

The enhanced model reproduces apoptosis after combined treatment with TNFα and the transcriptional inhibitor ActD as experimentally observed. The central players are ROS which are released from the mitochondria after TNFα treatment (v29). Normally, ROS are quickly scavenged by a protein P (v30) which is assumed to be a NF-κB target but whose identity could not yet been clarified definitely. However, after ActD treatment, P is not produced despite NF-κB activation by TNFα. In this case, ROS carry on their apoptotic impact by strongly increasing the amount of phosphorylated JNK (pJNK). In detail, ROS oxidise MKPs (v31) and thereby prevent them from dephosphorylating JNK. Additionally ROS are associated with a high increase of the overall amount of JNK (v33). According to the central role of ROS, the ROS scavenger BHA prevents apoptosis after TNFα plus ActD treatment in the model as it was also experimentally shown.

The apoptotic effect of pJNK after TNFα plus ActD treatment is realised in the model via the already discussed phosphorylation of Bim. In contrast to the first

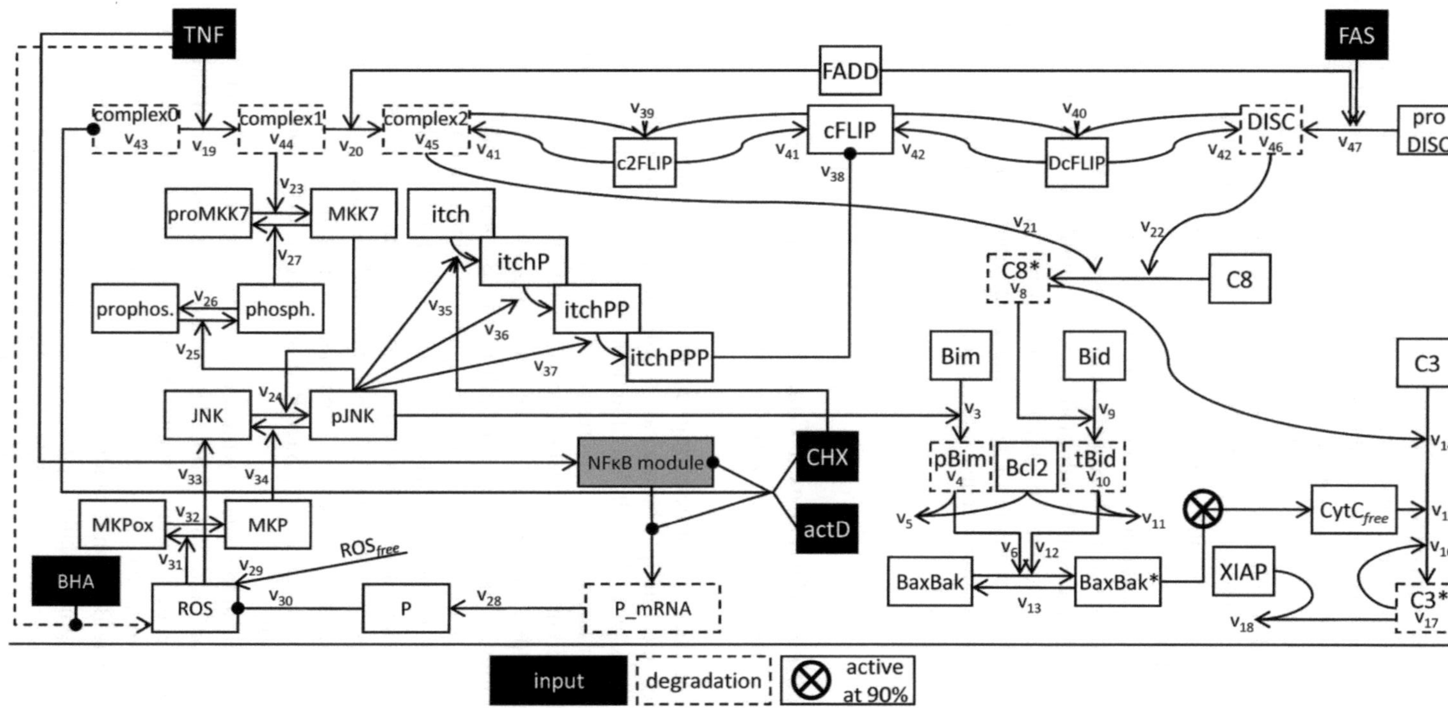

Fig. 5.3 Scheme of the enhanced apoptosis model, published in Schlatter et al. (2011)

sensitising model, a maximal amount of phosphorylated Bim which is able to induce cytochrome c release can be reached by the high amount of pJNK. Finally, an additional apoptotic route for phosphorylated JNK was included in the model which was reported by Chang et al. after apoptotic treatment with TNFα plus the translational blocker Cycloheximide (Chang et al. 2006). They showed that the amount of cFLIP is diminished via the E3 ligase Itch (v35–v38), which could however not be confirmed after ActD treatment. The reduced amount of cFLIP contributes to the apoptotic result as more caspase-8 can then be activated at complex II (v21).

In conclusion, the enhanced model is able to reproduce a variety of effects by the molecular interactions in a coherent and consistent way. The model originated from the connection of more simple sub-models. As the number of mathematical models of apoptosis and regeneration is constantly growing, the presented modeling process provides an example of how model integration of different submodels can produce more comprehensive models elucidating the behaviour of complex intracellular signaling.

5.3.2 *Boolean Models as a Tool for the Integration of Crosstalks Towards Apoptosis*

The large number of participating components in apoptosis, their complex dependencies and multiple biological stimuli make the model-based analysis of isolated network parts difficult and often less expressive. Unfortunately, the use of ODE models for larger networks is restricted due to limited quality and especially quantity of coherent biological data. The parameter identification for ODE models is mostly dependent on quantitative measurements which are still a bottleneck of systems biology. It is a special strength of the Boolean modeling (also called logical modeling) approach to reproduce qualitative interrelations. In comparison to kinetic models, Boolean models can be built and validated with a limited amount of experimental data. This makes logical modeling attractive for systems biology. Boolean modeling has become an important and frequently used approach in systems biology, which leads to important insights by model simulation and analysis (Schlatter et al. 2009b; Saez-Rodriguez et al. 2007; Samaga et al. 2009; Mai and Liu 2009).

Classical Boolean logic deals with two discrete states, e.g. “1” and “0” or “true” and “false” or “on” and “off” (Boole 1854), and was shown already in 1940 to be applicable to electrical relay circuits (Shannon 1940). Furthermore, biological signal flow networks can also be described by a logical approach (Thomas and D’Ari 1990). The Boolean formalism is especially useful for qualitative representation of signaling and regulatory networks where activation and inhibition are the essential processes (Thomas 1998). In a Boolean representation, the biological active state of a species can be translated into the “on” state, whereas the inactive state is represented by the "off" state. The species in a biological Boolean model can represent proteins, RNA and genes, for example. Enzymes play the role of switching other enzymes and genes "on" and "off". The knock-out of a certain gene as well as constitutive expression can be represented by fixing the node value of this species in the "off" or in the "on" state, respectively.

There are different interesting approaches for the simulation of Boolean networks. A description of so-called updating strategies allowing for simulating dynamic effects can be found in Faure et al. (2006). For our Boolean model we have used the approach of the logical steady state (LSS). A logical steady state is reached if the state of the model, i.e. the activity state of all species, does not change any more over time. Thus, the definition of LSS is independent from the chosen updating strategy and consequently the results are more universal. For a given input setting, all LSSs that can be reached from this state can be computed (Devloo et al. 2003). Here, an input setting is a set of initial values for certain nodes of the network. The input setting does not have to define an initial value for all nodes of the network. For the computation of LSSs for specified initial values, the software tool CellNetAnalyzer (CNA) is used (Klamt et al. 2006, 2007). It is available at http://www.mpi-magdeburg.mpg.de/projects/cna/cna.html. CNA also includes useful tools to predict and verify experimental data, examine the structure and the hierarchy of the model as well as the relevance of its components (Saez-Rodriguez et al. 2007; Klamt et al. 2006, 2007). Thereby it is easily operated via a graphical user interface.

A literature-based and in parts experimentally validated Boolean model of the central intrinsic and extrinsic apoptosis pathways as well as pathways connected with them was presented by the authors in Schlatter et al. (2009b). The model comprises 86 nodes and 125 interactions (Fig. 5.4). Descriptions of the network nodes as well as literature references to the equations can be found in Schlatter et al. (2009b). There are nine input nodes, namely glucagon, insulin, TNFα [TNF], Fas ligand [FasL], interleukin-1β [IL-1], UV-B irradiation [UV] and the following three special nodes:

The first special node represents Smac mimetics which are reagents that sensitise cells for apoptosis via the neutralisation of inhibitor of apoptosis proteins (IAPs) and can be employed to kill cancer cells (Wang et al. 2008; Li et al. 2004).

The second special input node "Type 2 receptor ligand" [T2RL] allows simulating apoptosis via the mitochondrial type II pathway in contrast to the type I pathway which proceeds via a direct activation of the caspase cascade (Scaffidi et al. 1998). The type I and type II pathways were shown to operate in the same cell type but under different culturing conditions, suggesting that cells are able to switch between both ways depending on external stimuli (Walter et al. 2008). However, the molecular mechanism of the switch itself has not yet been uncovered. Therefore, an additional node P representing some unknown protein or mechanism is introduced to model the switch behaviour.

Another speciality is the "housekeeping" input node, which is in the "on" state by default and shall reproduce constitutively expressed genes by fixing their node values in the "on" state (Fig. 5.4). However, the housekeeping node can also be used

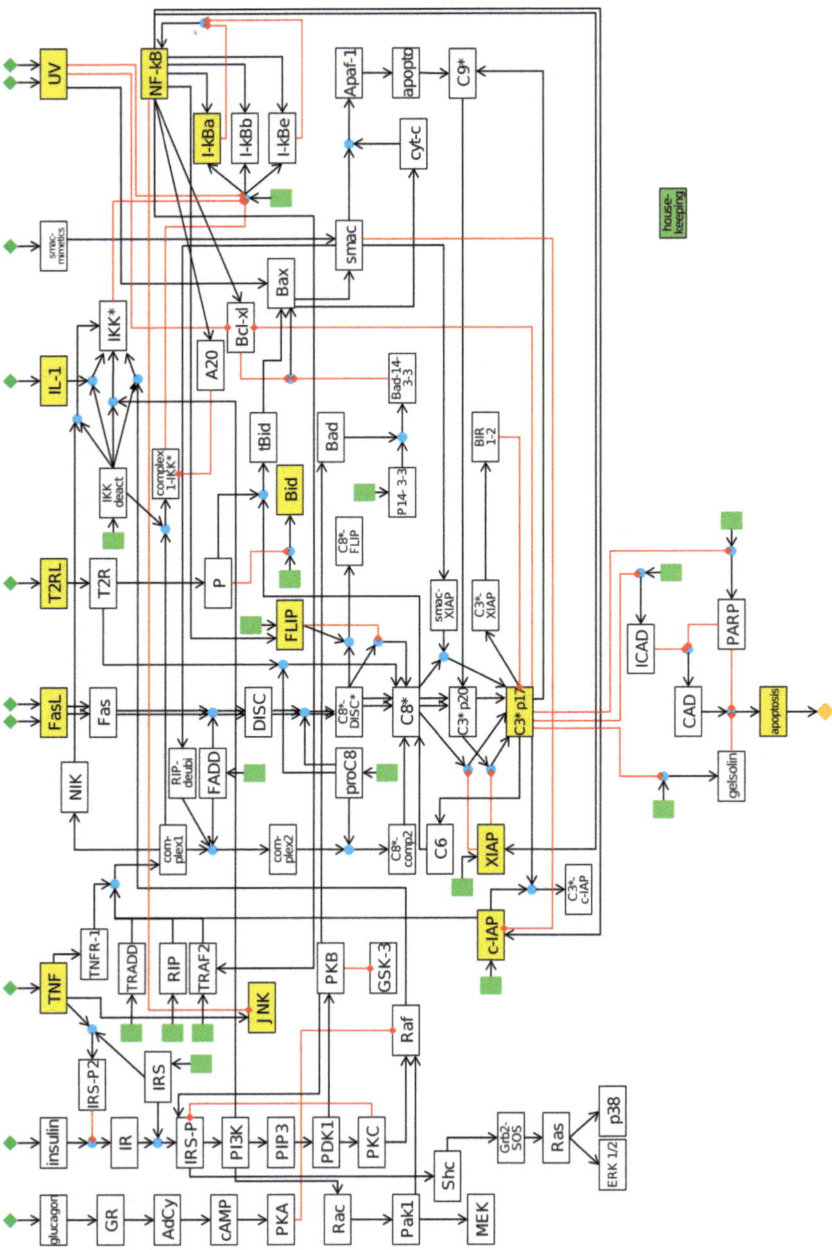

Fig. 5.4 Boolean apoptosis model, published in Schlatter et al. (2009b)

to simulate the influence of transcriptional inhibitors such as ActD or the translation inhibitor cycloheximide which are used frequently in experimental settings. In this case, the housekeeping node is manually set to the "off" state and in turn switches the according node values off.

Finally, the functional output node of the model is apoptosis. There are further output nodes in the model such as p38 or MEK which affect other proteins in the cell but not in the model. They are not followed-up in the model but are already tied for future model extensions.

The network map representing logical interactions is shown. The influence of the housekeeping node is depicted in green colour. Logical AND connections are represented by spheres. Activating arcs are represented by arrows and inhibiting arcs by lines with a bar. Figure adapted from Schlatter et al. (2009b).

Multi-value logic turned out to be indispensable to reproduce the behaviour of the apoptotic network (Schlatter et al. 2009b). Biochemical decisions are often made in increments caused by thresholds that are essential for setting boundaries between different states in living cells. This is especially true for apoptotic processes (Bentele et al. 2004; Legewie et al. 2006; Callus et al. 2008). Several biological settings could not be realised with single-value nodes. On that condition, the domain of some nodes had to be expanded to the values (0, 1, 2):

1. The FasL pathway was reported to show threshold behaviour (Lavrik et al. 2007; Bentele et al. 2004). Apoptosis is not reached in the model by FasL in activity state 1, denoted by [FasL (1)], but by FasL (2) reproducing the threshold behaviour of Fas signaling. However, FasL (1) activates several nodes in the network, and their influence and crosstalk with other signaling pathways can be analysed.
2. The nodes of the anti-apoptotic NF-κB regulated proteins c-FLIP, XIAP and c-IAPs (Kreuz et al. 2001; Stehlik et al. 1998) have graded effects in their "on" state. For example, caspase-3 p20 (2) can be further processed to the highly active caspase-3 p17 form which ensues in apoptosis if XIAP is low abundant as it is represented by XIAP (1). However, if XIAP is upregulated to value "2", it prevents processing and activation of caspase-3 p17.
3. Additionally, a multi-value node for UV irradiation was added based on own experimental results (Schlatter et al. 2009b). UV (1) leads to apoptosis, whereas UV (2) does not lead to apoptosis.

The logical apoptosis model is based on a vast number of different studies, which were performed in different organisms and were in part highly focusing on details (Schlatter et al. 2009b). Therefore, it was important to examine whether the behaviour emerging from these particular interactions in the model is coherent with experimental data on the behaviour of the whole network. Table 5.1 shows the model predictions for different proteins and stimuli which are critical for apoptosis represented by the resulting logical steady state values of the model. The model values of the input nodes are given in parentheses and mock is represented by the logical steady state of the model without activation of any input node.

Table 5.1 Central results of the logical apoptosis model that have been experimentally validated

		NF-κB	IκB-α	c-IAP	XIAP	C3*p17	FLIP	Bid	JNK	Apoptosis
Cell type	Stimulation	EMSA	Western	qRT-PCR	Western	DEVD	qRT-PCR	Western	Western	MTT
Jurkats	Mock	0	1	1	1	0	1	1	0	0
Jurkats	T2RL (1)	0	1	1	1	2	1	0	0	1
Hepatocytes	Mock	0	1	1	1	0	1	1	0	0
Hepatocytes	FasL (2)	0	1	1	1	2	1	1	0	1
Hepatocytes	TNF (1)	1	0	2	2	0	2	1	1	0
Hepatocytes	IL-1 (1)	1	0	2	2	0	2	1	0	0
Hepatocytes	UV (1)	0	1	1	1	2	1	1	0	1
Hepatocytes	UV (2)	1	0	2	2	0	2	1	0	0

All model predictions listed in Table 5.1 are in accordance with measurement data. They were achieved on the first attempt without changing the model, apart from the effect of UV irradiation on the network. This shows that the known molecular interactions consistently explain the behaviour of the network.

In the experiments, two different cell types were used to account for the distinct signaling mechanisms in cells using the type I (primary mouse hepatocytes treated with FasL) and the type II (human Jurkat T cells treated with FasL representing the T2RL node) apoptotic pathways. Details on the experimental procedures as well as figures presenting the results can be found in Schlatter et al. (2009b).

During experimental validation of the model, a dose-dependent NF-κB activation and apoptosis after UV irradiation were found to occur in primary mouse hepatocytes. As the Boolean model does not reflect quantitative units, different strengths of UV irradiation have been tested for the model validation. Based on the results, two distinct levels for the UV input node were implemented. UV (1) represents the stimulation of mouse hepatocytes with 300 J/m^2 UV irradiation and UV (2) with 600 J/m^2. Weak UV irradiation leads to weak NF-κB activation and no c-IAP2 and c-FLIP mRNA upregulation. As there is no signaling effect on the subsequent nodes, the model shows NF-κB (0) in this setting. As a consequence, mouse hepatocytes show significantly increased caspase-3 activity and consequently apoptosis can be observed as expected after UV irradiation. In contrast, the higher dose of UV irradiation leads to strong NF-κB activation, and subsequently c-IAP2 and c-FLIP mRNA are upregulated. This correlates with previous findings showing a marked NF-κB induction after strong translational inhibition and relative resistance to lower doses (O'Dea et al. 2008). The proteins c-IAP2 and c-FLIP function as anti-apoptotic inhibitors and prevent caspase-3 activation in this setting. Accordingly, cells show less cytotoxicity after strong UV irradiation. The updated model version reflects the network behaviour in response to UV irradiation.

It is impossible to test every signaling scenario of the presented apoptosis model due to technical limitations and the mere number of nodes. However, the consistency of the model with the performed validation experiments indicates that the model is basically correct and has explanatory power.

Analysis of the model revealed a tight regulation emerging from high connectivity and spanning crosstalks and a particular importance of feedback loops. Detailed analysis can be found in Schlatter et al. (2009b). One of the results consists of predictions for double stimulation scenarios. Therefore, the resulting value for the apoptosis node has been systematically simulated and listed in Table 5.2. The diagonal shows the resulting apoptosis value for the according single stimulations. One would assume the outcome for two combined stimuli to follow the rules $0 + 0 = 0$, $1 + 1 = 1$ and $0 + 1 = 1$. However, there are some aberrations which are highlighted bold in Table 5.2. There are some interesting crosstalks concerning UV stimulation. The anti-apoptotic effects of insulin and IL-1β also prevent apoptosis in combination with UV (1). However, in combination with TNFα, apoptosis is still enforced by UV (1) as Smac is released according to the model by UV irradiation and counteracts XIAP upregulation. The input combinations of UV (2) with TNF and FasL (1) also lead to apoptosis as the latter activate caspase-8 (1).

Table 5.2 Apoptosis node value for all double stimulation scenarios of the model

	Glucagon	Insulin	TNF	FasL (1)	FasL (2)	T2RL	IL-1	smac-mimetics	UV (1)	UV (2)
Glucagon	0	0	0	0	1	1	0	0	1	0
Insulin		0	0	0	**0**	1	0	0	**0**	0
TNF			0	0	**0**	1	0	**1**	1	**1**
FasL (1)				0	–	1	0	**1**	1	**1**
FasL (2)					1	1	**0**	1	1	**0**
T2RL						1	1	1	1	1
IL-1							0	0	**0**	0
Smac-mimetics								0	1	0
UV (1)									1	–
UV (2)										0

In contrast, the combination of FasL (2) and UV (2) does not cause apoptosis in the model as the NF-κB activation by UV (2) is dominant in this setting.

Additionally, Smac-mimetics lead to apoptosis in combination with FasL (1). IL-1β, insulin and TNFα prevent apoptosis after FasL (2) stimulation in the model. However, our latest studies showed that the concurrent stimulation with TNFα and FasL does not prevent, but still induces apoptosis in hepatocytes. Moreover, the sensitising effect, as described in Sect. 5.3.1, is also currently not covered by the Boolean model. It can be expected that the extension of the Boolean model considering this sensitising effect and including the influence of JNK via Bim on Bax/Bak may in turn provide new predictions for additional crosstalks on a dynamical level in the future.

5.4 Conclusion

Here, we present two complementary approaches of modeling death receptor-mediated cell death in hepatocytes. Kinetic modeling is suited to provide a detailed, quantitative and dynamic explanation for the different apoptosis decisions under varying scenarios triggered by two interacting signaling pathways. Boolean modeling is able to describe larger networks on a qualitative level. We focused on a comprehensive and coherent description of the apoptosis decision after different stimuli.

The use of ordinary differential equation (ODE) models is a standard approach in systems biology allowing the description of dynamic behaviour over time. In the presented work, an ODE model of TNFα and FasL sensitising was extended by adding the regulation of pJNK and the generation of ROS after combined TNFα and ActD treatment and their impact on the signaling network. Moreover, a published NF-κB model was integrated in the framework. The qualitative interactions in the network are based on literature and measurement data for murine hepatocytes which clearly document the modelled functional mechanisms.

The experimental approach combined with the extended model provides a deep insight into TNF signaling in murine hepatocytes. Systems biology has just been passing from dynamic modeling of single pathways to dynamic modeling of comprehensive and intensively cross-linked networks. As already demonstrated with other cells, more information on apoptosis and NF-κB can be gained by adding further levels of detail in existing models (Cheong et al. 2008; Lavrik et al. 2009; Lavrik 2010). We here provide an example for the integration of several dynamical mathematical models into a holistic model describing complex crosstalks. The presented model can be used for the integration of further pathway models. Since subtle dependencies between different pathways exist, the integration of additional pathways will probably improve the analysis of all pathways involved and the understanding of the holistic network.

Boolean modeling employing also multi-value logic allows the reproduction of the apoptotic signaling network in a comprehensive way. The presented apoptosis model can be used for comparison with own results as well as for further analyses. It can be modified and expanded to other cell types, additional pathways or crosstalks. The integration of numerous signaling pathways into one model can especially be used to predict the systems behaviour after multiple stimuli. In particular, any kind of knock-out or knock-in scenario can be easily simulated with the model by setting certain nodes or interactions to the desired value. Subsequently, resulting variations in signaling behaviour and the changed network topology can be analysed. On the other hand one can search for minimal intervention sets (Klamt and Gilles 2004). Thereby the algorithm computes all possibilities to reach a user-defined network state under user-defined constraints as fixed states or maximum number of interventions. Finally, uncovering sensitive points in the network and failure modes of the system concerning specific questions will provide suggestions for biological experimental design as well as predictions on how the system reacts in response to selected challenges. Beyond this functionality a comprehensive Boolean model can help to define a meaningful system boundary for an ODE model of an included single pathway. The connectivity of sub-networks and single components via crosstalks is helpful to include all essential interactions when focusing on a smaller subsystem or specific question.

In the future, modeling of apoptosis in hepatocytes will proceed on different routes on the way to a functional representation of the whole liver. On the one hand, there are still functional properties of cell signaling to elucidate. A critical point thereby is the switch between type I and II apoptosis which is dependent on culturing conditions, suggesting a central role of integrin signaling (Walter et al. 2008). Additionally, the detailed steps of mitochondrial pore formation are not yet fully understood and usually still modelled in a more phenomenological than mechanistic way. Beyond the investigation of single signaling pathways, the combination of the according mathematical models will be necessary to understand complex crosstalks and interactions not only inside the hepatocytes but also between the different cells in the liver. ODE as well as Boolean models will contribute to this process. As inflammation processes are of central interest regarding apoptosis in the liver from a medical and clinical point of view, especially cell–cell models describing the interactions between hepatocytes and cells of the immune system will be of increasing importance and a next step towards a mathematical liver model supporting clinical research.

Acknowledgements This work was funded by the BMBF (German Federal Ministry of Education and Research) within the Virtual Liver Network (http://www.virtual-liver.de).

References

Akazawa Y, Gores GJ (2007) Death receptor-mediated liver injury. Semin Liver Dis 27:327–338
Baud V, Karin M (2001) Signal transduction by tumor necrosis factor and its relatives. Trends Cell Biol 11:372–377
Baud V, Karin M (2009) Is NF-kappaB a good target for cancer therapy? Hopes and pitfalls. Nat Rev Drug Discov 8:33–40
Ben MT, Barash H, Kang TB, Kim JC, Kovalenko A, Gross E, Schuchmann M, Abramovitch R, Galun E, Wallach D (2007) Role of caspase-8 in hepatocyte response to infection and injury in mice. Hepatology 45:1014–1024
Bentele M, Lavrik I, Ulrich M, Stosser S, Heermann DW, Kalthoff H, Krammer PH, Eils R (2004) Mathematical modeling reveals threshold mechanism in CD95-induced apoptosis. J Cell Biol 166:839–851
Boole G (1854) An investigation of the laws of thought on which are founded the mathematical theories of logic and probabilities. Walter and Maberly, London
Bradham CA, Plumpe J, Manns MP, Brenner DA, Trautwein C (1998) Mechanisms of hepatic toxicity. I. TNF-induced liver injury. Am J Physiol 275:G387–G392
Callus BA, Moujallad DM, Silke J, Gerl R, Jabbour AM, Ekert PG, Vaux DL (2008) Triggering of apoptosis by Puma is determined by the threshold set by prosurvival Bcl-2 family proteins. J Mol Biol 384:313–323
Calzone L, Tournier L, Fourquet S, Thieffry D, Zhivotovsky B, Barillot E, Zinovyev A (2010) Mathematical modelling of cell-fate decision in response to death receptor engagement. PLoS Comput Biol 6:e1000702
Canbay A, Friedman S, Gores GJ (2004) Apoptosis: the nexus of liver injury and fibrosis. Hepatology 39:273–278
Chang L, Kamata H, Solinas G, Luo JL, Maeda S, Venuprasad K, Liu YC, Karin M (2006) The E3 ubiquitin ligase itch couples JNK activation to TNFalpha-induced cell death by inducing c-FLIP(L) turnover. Cell 124:601–613
Cheong R, Hoffmann A, Levchenko A (2008) Understanding NF-kappaB signaling via mathematical modeling. Mol Syst Biol 4:192
Corazza N, Jakob S, Schaer C, Frese S, Keogh A, Stroka D, Kassahn D, Torgler R, Mueller C, Schneider P, Brunner T (2006) TRAIL receptor-mediated JNK activation and Bim phosphorylation critically regulate Fas-mediated liver damage and lethality. J Clin Invest 116:2493–2499
Costelli P, Aoki P, Zingaro B, Carbo N, Reffo P, Lopez-Soriano FJ, Bonelli G, Argiles JM, Baccino FM (2003) Mice lacking TNFalpha receptors 1 and 2 are resistant to death and fulminant liver injury induced by agonistic anti-Fas antibody. Cell Death Differ 10:997–1004
Deng Y, Ren X, Yang L, Lin Y, Wu X (2003) A JNK-dependent pathway is required for TNFalpha-induced apoptosis. Cell 115:61–70
Devloo V, Hansen P, Labbe M (2003) Identification of all steady states in large networks by logical analysis. Bull Math Biol 65:1025–1051
Faure A, Naldi A, Chaouiya C, Thieffry D (2006) Dynamical analysis of a generic Boolean model for the control of the mammalian cell cycle. Bioinformatics 22:e124–e131
Fausto N, Campbell JS, Riehle KJ (2006) Liver regeneration. Hepatology 43:S45–S53

Hatano E (2007) Tumor necrosis factor signaling in hepatocyte apoptosis. J Gastroenterol Hepatol 22(Suppl 1):S43–S44

Hughes MA, Harper N, Butterworth M, Cain K, Cohen GM, MacFarlane M (2009) Reconstitution of the death-inducing signaling complex reveals a substrate switch that determines CD95-mediated death or survival. Mol Cell 35:265–279

Kamata H, Honda S, Maeda S, Chang L, Hirata H, Karin M (2005) Reactive oxygen species promote TNFalpha-induced death and sustained JNK activation by inhibiting MAP kinase phosphatases. Cell 120:649–661

Karin M (2006) Nuclear factor-kappaB in cancer development and progression. Nature 441:431–436

Karin M, Lin A (2002) NF-κB at the crossroads of life and death. Nat Immunol 3:221–227

Kaufmann T, Tai L, Ekert PG, Huang DC, Norris F, Lindemann RK, Johnstone RW, Dixit VM, Strasser A (2007) The BH3-only protein bid is dispensable for DNA damage- and replicative stress-induced apoptosis or cell-cycle arrest. Cell 129:423–433

Kaufmann T, Jost PJ, Pellegrini M, Puthalakath H, Gugasyan R, Gerondakis S, Cretney E, Smyth MJ, Silke J, Hakem R, Bouillet P, Mak TW, Dixit VM, Strasser A (2009) Fatal hepatitis mediated by tumor necrosis factor TNFalpha requires caspase-8 and involves the BH3-only proteins Bid and Bim. Immunity 30:56–66

Klamt S, Gilles ED (2004) Minimal cut sets in biochemical reaction networks. Bioinformatics 20:226–234

Klamt S, Saez-Rodriguez J, Lindquist JA, Simeoni L, Gilles ED (2006) A methodology for the structural and functional analysis of signaling and regulatory networks. BMC Bioinformatics 7:56

Klamt S, Saez-Rodriguez J, Gilles ED (2007) Structural and functional analysis of cellular networks with cell netanalyzer. BMC Syst Biol 1:2

Klingmuller U, Bauer A, Bohl S, Nickel PJ, Breitkopf K, Dooley S, Zellmer S, Kern C, Merfort I, Sparna T, Donauer J, Walz G, Geyer M, Kreutz C, Hermes M, Gotschel F, Hecht A, Walter D, Egger L, Neubert K, Borner C, Brulport M, Schormann W, Sauer C, Baumann F, Preiss R, MacNelly S, Godoy P, Wiercinska E, Ciuclan L, Edelmann J, Zeilinger K, Heinrich M, Zanger UM, Gebhardt R, Maiwald T, Heinrich R, Timmer J, von WF, Hengstler JG (2006) Primary mouse hepatocytes for systems biology approaches: a standardized in vitro system for modelling of signal transduction pathways. Syst Biol 153:433–447

Krammer PH (2000) CD95's deadly mission in the immune system. Nature 407:789–795

Kreuz S, Siegmund D, Scheurich P, Wajant H (2001) NF-κB inducers upregulate cFLIP, a cycloheximide-sensitive inhibitor of death receptor signaling. Mol Cell Biol 21:3964–3973

Kreuz S, Siegmund D, Rumpf JJ, Samel D, Leverkus M, Janssen O, Hacker G, Dittrich-Breiholz O, Kracht M, Scheurich P, Wajant H (2004) NFkappaB activation by Fas is mediated through FADD, caspase-8, and RIP and is inhibited by FLIP. J Cell Biol 166:369–380

Lavrik IN (2010) Systems biology of apoptosis signaling networks. Curr Opin Biotechnol 21:551–555

Lavrik IN, Golks A, Riess D, Bentele M, Eils R, Krammer PH (2007) Analysis of CD95 threshold signaling: triggering of CD95 (FAS/APO-1) at low concentrations primarily results in survival signaling. J Biol Chem 282:13664–13671

Lavrik IN, Eils R, Fricker N, Pforr C, Krammer PH (2009) Understanding apoptosis by systems biology approaches. Mol Biosyst 5:1105–1111

Legewie S, Bluthgen N, Herzel H (2006) Mathematical modeling identifies inhibitors of apoptosis as mediators of positive feedback and bistability. PLoS Comput Biol 2:e120

Li L, Thomas RM, Suzuki H, De Brabander JK, Wang X, Harran PG (2004) A small molecule Smac mimic potentiates TRAIL- and TNF-α mediated cell death. Science 305:1471–1474

Lin A (2003) Activation of the JNK signaling pathway: breaking the brake on apoptosis. Bioessays 25:17–24

Lipniacki T, Paszek P, Brasier AR, Luxon B, Kimmel M (2004) Mathematical model of NF-κB regulatory module. J Theor Biol 228:195–215

Li-Weber M, Krammer PH (2003) Function and regulation of the CD95 (APO-1/Fas) ligand in the immune system. Semin Immunol 15:145–157

Mai Z, Liu H (2009) Boolean network-based analysis of the apoptosis network: irreversible apoptosis and stable surviving. J Theor Biol 259:760–769

Malhi H, Gores GJ (2008) Cellular and molecular mechanisms of liver injury. Gastroenterology 134:1641–1654

Malhi H, Guicciardi ME, Gores GJ (2010) Hepatocyte death: a clear and present danger. Physiol Rev 90:1165–1194

McKenzie MD, Carrington EM, Kaufmann T, Strasser A, Huang DC, Kay TW, Allison J, Thomas HE (2008) Proapoptotic BH3-only protein Bid is essential for death receptor-induced apoptosis of pancreatic β-cells. Diabetes 57:1284–1292

Michalopoulos GK (2007) Liver regeneration. J Cell Physiol 213:286–300

Micheau O, Tschopp J (2003) Induction of TNF receptor I-mediated apoptosis via two sequential signaling complexes. Cell 114:181–190

Micheau O, Lens S, Gaide O, Alevizopoulos K, Tschopp J (2001) NF-κB signals induce the expression of c-FLIP. Mol Cell Biol 21:5299–5305

Mohammed FF, Khokha R (2005) Thinking outside the cell: proteases regulate hepatocyte division. Trends Cell Biol 15:555–563

Nakano H, Nakajima A, Sakon-Komazawa S, Piao JH, Xue X, Okumura K (2006) Reactive oxygen species mediate crosstalk between NF-κB and JNK. Cell Death Differ 13:730–737

Neumann L, Pforr C, Beaudouin J, Pappa A, Fricker N, Krammer PH, Lavrik IN, Eils R (2010) Dynamics within the CD95 death-inducing signaling complex decide life and death of cells. Mol Syst Biol 6:352

Nowak M, Gaines GC, Rosenberg J, Minter R, Bahjat FR, Rectenwald J, MacKay SL, Edwards CK III, Moldawer LL (2000) LPS-induced liver injury in D-galactosamine-sensitized mice requires secreted TNF-α and the TNF-p55 receptor. Am J Physiol Regul Integr Comp Physiol 278:R1202–R1209

O'Dea EL, Kearns JD, Hoffmann A (2008) UV as an amplifier rather than inducer of NF-κB activity. Mol Cell 30:632–641

Ogasawara J, Watanabe-Fukunaga R, Adachi M, Matsuzawa A, Kasugai T, Kitamura Y, Itoh N, Suda T, Nagata S (1993) Lethal effect of the anti-Fas antibody in mice. Nature 364:806–809

Papa S, Bubici C, Zazzeroni F, Pham CG, Kuntzen C, Knabb JR, Dean K, Franzoso G (2006) The NF-κB-mediated control of the JNK cascade in the antagonism of programmed cell death in health and disease. Cell Death Differ 13:712–729

Papa S, Bubici C, Zazzeroni F, Franzoso G (2009) Mechanisms of liver disease: cross-talk between the NF-κB and JNK pathways. Biol Chem 390:965–976

Peter ME, Budd RC, Desbarats J, Hedrick SM, Hueber AO, Newell MK, Owen LB, Pope RM, Tschopp J, Wajant H, Wallach D, Wiltrout RH, Zornig M, Lynch DH (2007) The CD95 receptor: apoptosis revisited. Cell 129:447–450

Philippi N, Walter D, Schlatter R, Ferreira K, Ederer M, Sawodny O, Timmer J, Borner C, Dandekar T (2009) Modeling system states in liver cells: survival, apoptosis and their modifications in response to viral infection. BMC Syst Biol 3:97

Plati J, Bucur O, Khosravi-Far R (2011) Apoptotic cell signaling in cancer progression and therapy. Integr Biol 3:279–296

Saez-Rodriguez J, Simeoni L, Lindquist JA, Hemenway R, Bommhardt U, Arndt B, Haus UU, Weismantel R, Gilles ED, Klamt S, Schraven B (2007) A logical model provides insights into T cell receptor signaling. PLoS Comput Biol 3:e163

Samaga R, Saez-Rodriguez J, Alexopoulos LG, Sorger PK, Klamt S (2009) The logic of EGFR/ErbB signaling: theoretical properties and analysis of high-throughput data. PLoS Comput Biol 5:e1000438

Scaffidi C, Fulda S, Srinivasan A, Friesen C, Li F, Tomaselli KJ, Debatin KM, Krammer PH, Peter ME (1998) Two CD95 (APO-1/Fas) signaling pathways. EMBO J 17:1675–1687

Schlatter R, Conzelmann H, Gilles ED, Sawodny O, Sauter T (2009a) Analysis of an apoptotic core model focused on experimental design using artificial data. IET Syst Biol 3:255–265

Schlatter R, Schmich K, Avalos VI, Scheurich P, Sauter T, Borner C, Ederer M, Merfort I, Sawodny O (2009b) ON/OFF and beyond-a boolean model of apoptosis. PLoS Comput Biol 5:e1000595

Schlatter R, Schmich K, Trefzger J, Sawodny O, Ederer M, Merfort I (2011) Modeling of TNFalpha-induced apoptosis in hepatocytes. PLoS One 6:e18646

Schmich K, Schlatter R, Corazza N, Ferreira KS, Ederer M, Brunner T, Borner C, Merfort I (2011) Tumor necrosis factor alpha sensitizes primary murine hepatocytes to Fas/CD95-induced apoptosis in a Bim- and Bid-dependent manner. Hepatology 53:282–292

Schungel S, Buitrago-Molina LE, Nalapareddy P, Lebofsky M, Manns MP, Jaeschke H, Gross A, Vogel A (2009) The strength of the Fas ligand signal determines whether hepatocytes act as type 1 or type 2 cells in murine livers. Hepatology 50:1558–1566

Shannon CA (1940) Symbolic analysis of relay and switching circuits. Massachusetts Institute of Technology, Department of Electrical Engineering, Cambridge

Stehlik C, De MR, Kumabashiri I, Schmid JA, Binder BR, Lipp J (1998) Nuclear factor (NF)-kappaB-regulated X-chromosome-linked iap gene expression protects endothelial cells from tumor necrosis factor alpha-induced apoptosis. J Exp Med 188:211–216

Strasser A, Jost PJ, Nagata S (2009) The many roles of FAS receptor signaling in the immune system. Immunity 30:180–192

Sun B, Karin M (2008) NF-κB signaling, liver disease and hepatoprotective agents. Oncogene 27:6228–6244

Taub R (2004) Liver regeneration: from myth to mechanism. Nat Rev Mol Cell Biol 5:836–847

Thomas R (1998) Laws for the dynamics of regulatory networks. Int J Dev Biol 42:479–485

Thomas R, D'Ari R (1990) Biological feedback. CRC, Boca Raton

Varfolomeev EE, Ashkenazi A (2004) Tumor necrosis factor: an apoptosis JuNKie? Cell 116:491–497

Vince JE, Wong WW, Khan N, Feltham R, Chau D, Ahmed AU, Benetatos CA, Chunduru SK, Condon SM, McKinlay M, Brink R, Leverkus M, Tergaonkar V, Schneider P, Callus BA, Koentgen F, Vaux DL, Silke J (2007) IAP antagonists target cIAP1 to induce TNFalpha-dependent apoptosis. Cell 131:682–693

Wajant H, Pfizenmaier K, Scheurich P (2003) Tumor necrosis factor signaling. Cell Death Differ 10:45–65

Walter D, Schmich K, Vogel S, Pick R, Kaufmann T, Hochmuth FC, Haber A, Neubert K, McNelly S, Von WF, Merfort I, Maurer U, Strasser A, Borner C (2008) Switch from type II to I Fas/CD95 death signaling on in vitro culturing of primary hepatocytes. Hepatology 48:1942–1953

Wang L, Du F, Wang X (2008) TNF-α induces two distinct caspase-8 activation pathways. Cell 133:693–703

Yazdanpanah B, Wiegmann K, Tchikov V, Krut O, Pongratz C, Schramm M, Kleinridders A, Wunderlich T, Kashkar H, Utermohlen O, Bruning JC, Schutze S, Kronke M (2009) Riboflavin kinase couples TNF receptor 1 to NADPH oxidase. Nature 460:1159–1163

Yin XM, Wang K, Gross A, Zhao Y, Zinkel S, Klocke B, Roth KA, Korsmeyer SJ (1999) Bid-deficient mice are resistant to Fas-induced hepatocellular apoptosis. Nature 400:886–891

Chapter 6
Understanding Different Types of Cell Death Using Systems Biology

Laurence Calzone, Andrei Zinovyev, and Boris Zhivotovsky

Abstract This chapter is devoted to the mathematical modeling of cellular decisions between death and life (referred to as cell fate decision). These decisions determine many cell events in multicellular and unicellular organisms. Understanding the principles of cell fate decisions is crucial for the studies of some tissue functioning (such as gut epithelium) and for the comprehension of tumour development, for which the tightly regulated mechanism of balancing between survival and death is violated towards survival. In a broader context, cell fate decisions are examples of cellular decision-making mechanisms which are abundant in all living organisms from viruses and bacteria to mammals (Balazsi et al. 2011).

6.1 Problem Statement: Modeling Cell Fate Decision Process

This chapter is devoted to the mathematical modeling of cellular decisions between death and life (referred to as cell fate decision). These decisions determine many cell events in multicellular and unicellular organisms. Understanding the principles of cell fate decisions is crucial for the studies of some tissue functioning (such as gut epithelium) and for the comprehension of tumour development, for which the tightly regulated mechanism of balancing between survival and death is violated towards survival. In a broader context, cell fate decisions are examples of cellular

L. Calzone • A. Zinovyev
Institut Curie, Paris, France

Ecole des Mines ParisTech, Paris, France

INSERM U900, Paris, France

B. Zhivotovsky (✉)
Institute of Environmental Medicine, Karolinska Institutet, Box 210, SE-17177 Stockholm, Sweden
e-mail: Boris.Zhivotovsky@ki.se

I.N. Lavrik (ed.), *Systems Biology of Apoptosis*, DOI 10.1007/978-1-4614-4009-3_6,

decision-making mechanisms which are abundant in all living organisms from viruses and bacteria to mammals (Balazsi et al. 2011).

Cell fate decision process clearly possesses characteristics of complex systems such that the decision outcome can be rarely attributed to the activity of one single molecular player. The mechanisms underlying these decisions have systemic properties of complex genetic networks. In 1957, Waddington already proposed the idea of an epigenetic landscape in the context of morphogenesis (Waddington 1957). According to him, the cell develops through a series of irreversible transitions that are controlled by complex networks of interacting genes. These networks are indeed determinant, but there are more parameters intervening in the choice of the cellular outcome. It is important to note that the roles of external noise, internal noise and the properties of the surrounding environment are crucial in these cellular choices. Based on the idea of studying the dynamical properties of a complex system of interacting genes, a systems biology approach seems to be suited in deciphering the cell fate decision process.

The term "programmed cell death" was coined in 1965 (Lockshin and Williams 1965) and constitutes one of the three possible cell fates, apart from proliferation and differentiation, and is one of the elementary processes in cell biology. In the normal, healthy human adult, cell death affects several millions of cells per second. Deregulation of apoptosis is involved in several pathologies. Kerr, Wyllie and Currie (1972) described cell death via apoptosis, which is an example of programmed cell death, as a basic biological phenomenon with wide-ranging implications in tissue kinetics. This influential paper implicated apoptosis not only in normal cell turnover, but also in the spontaneous elimination of potentially dangerous cells and, hence, in the pathogenesis of cancer. Moreover, the authors speculated that the mechanism(s) leading to spontaneous apoptosis in growing malignant neoplasms might also be applicable in therapeutically induced tumour regression. Today, it is generally agreed that cell populations are tightly regulated with regard to proliferation, differentiation and death. Dysfunction of any one of these processes can result in either uncontrolled cell growth or uncontrolled cell death. Yet, it remains unclear to what extent a perturbation of cell proliferation and/or cell death regulation must occur in order to influence the pathogenesis of various diseases. In their 1972 paper, Kerr and colleagues compared two modes of cell death, apoptosis and necrosis, suggesting that the former represents an example of gene-regulated processes, while the latter is simply a passive event. Since then, the field of cell death has become an increasingly important area of biomedical research.

The apoptotic dying cell undergoes rapid changes, which are reflected in both its structure and biochemistry. Morphologically, apoptosis is characterised by margination and condensation of nuclear chromatin (pyknosis), cell shrinkage, nuclear fragmentation and blebbing of the plasma membrane. The cell subsequently breaks up into membrane-enclosed fragments, termed apoptotic bodies, which are rapidly recognised and engulfed by neighbouring cells or macrophages. Considerable biochemical changes occur within the apoptotic cell to facilitate neat packaging and removal of the apoptotic bodies by phagocytosis.

Necrosis has long been regarded as the result of an accidental and uncontrolled process, usually caused by external factors to the cell or tissue, such as infection, toxins, heat or trauma. It is characterised by disruption of the plasma membrane, and of the membranes of intracellular organelles, cell swelling, chromatin digestion, DNA hydrolysis and, finally, cell lysis. Necrosis is often associated with local inflammation, triggered by the release of factors from dead cells that alert the innate immune system. Necrosis is known to play a prominent role in many pathological conditions, including ischaemia/reperfusion (e.g. stroke and myocardial infarction), trauma and some forms of neurodegeneration.

Accumulating evidence suggests that necrotic cell death might also be regulated by a specific set of signal transduction pathways and degradative mechanisms (Golstein and Kroemer 2007). Similar to apoptosis, cell death with a necrotic appearance can contribute to embryonic development and adult tissue homeostasis. At present the Nomenclature Committee of Cell Death has recognised several modes of cells death (Kroemer et al. 2009), which are divided into typical and atypical. Although these modes are all characterised by distinct morphologies, the molecular mechanisms underlying a majority of these cell death modalities have not been sufficiently investigated. The modeling of various cell death modalities and crosstalk between them will help in the understanding of their regulation pathways.

6.2 Conceptual Description of Biological Switches

Cellular decision is conceptually linked to the notion of cellular switches. These switches are molecular devices that integrate a variety of external signals and end up in a particular state. Cellular switches can be represented by single molecules. In this case, they are called molecular switches and they are governed by the state of one single component. However, in cellular decision processes, the switch is frequently an emergent property of an interaction network. Even if some of the network components can play a more important role in determining the outcome of the switch, these players are part of a bigger network that controls the final choice.

Until recently, a requirement for gene expression was documented only for apoptotic and autophagic cell death. Interestingly, certain genes and their products, e.g. p53, Bcl-2 family proteins, calpain, etc., are important for both these modes of cell death. Recent work indicates that basal p53 activity suppresses autophagy, whereas the activation of p53 by some stimuli induces autophagy as well as apoptosis mediated by the Bcl-2 family proteins, PUMA and NOXA. As mentioned above, for many years, necrosis was regarded as the result of an accidental and uncontrolled process, but many observations revealed that necrotic cell death might also be mediated by a specific set of signal transduction pathways and degradative mechanisms (Golstein and Kroemer 2007). Some gene products, such as TNFR, CD95, TRAIL-R and RIP1, might trigger both apoptosis and necrosis, depending on their interaction with other proteins. Moreover, there exist some crosstalks between these two cell death modalities. For example, inactivation of caspases might cause a shift from apoptosis to necrosis, or to a mixture of these two cell death modes. Recently, the term necroptosis has been introduced to designate a special type of programmed necrosis that depends on serine/threonine kinase RIP1 activity (Degterev et al. 2008). In a genome-wide siRNA screen for regulators of necroptosis, a set of 432 genes was identified. Among these genes was a subset of 32 genes that act downstream of, and/or as regulators of RIP1 kinase, and 7 genes involved in both necroptosis and apoptosis. Interestingly, Bmf, a BH3-only, Bcl-2 family member, was shown to be required for death receptor-induced necroptosis. In fact, it seems that Bcl-2 family proteins are essential for the regulation of a majority of programmed cell death modalities.

Depending on the type of lethal stimulus, the cell death process can be initiated in different intracellular compartments, and crosstalk between these compartments is essential for all cell death modalities. Moreover, it seems that various organelles might trigger cell death by specific stress sensors and transmit cell death modulating signals throughout the cell. This inter-organelle crosstalk apparently involves several molecular "switches" within the signaling network. Thus, p53 can be activated in response to DNA damage, or to changes in redox balance in the mitochondria, and Bcl-2 family proteins might act at the level of the mitochondria, ER or nucleus. Nuclear p53 promotes the transcription of pro-apoptotic and cell cycle-arresting genes, and also can act as an autophagy-inducing transcription factor. In contrast, cytoplasmic p53 might trigger apoptosis and exert an autophagy-inhibitory function, although the precise molecular mechanism(s) of this dual function is (are) not completely known. Another example of crosstalk between apoptosis and autophagy was described recently (Wirawan et al. 2010). Beclin-1 was originally identified as a Bcl-2-interacting protein, whose autophagic function can be inhibited by both Bcl-2 and Bcl-XL. Notably, although Beclin-1 possesses a BH3-only domain, and all BH3-only proteins of the Bcl-2 family are well-known inducers of apoptosis, Beclin-1 fails to induce apoptosis and by stimulation of autophagy offers cytoprotection against various apoptosis triggers. However, in response to growth factor withdrawal, when autophagy precedes apoptosis, caspase-mediated cleavage of Beclin-1 inactivates autophagy and stimulates apoptosis by promoting the release of pro-apoptotic factors from mitochondria. It seems that in this experimental model, a caspase-generated fragment of Beclin-1 triggers an amplifying loop enhancing apoptosis.

Importantly, depending on the nature and severity of the stimulus, and on the cell type, the hierarchy of inter-organelle crosstalk might result in different cell death modalities. Moreover, in some cases, suppression of the function of a particular intracellular compartment might switch one mode of cell death to another. For example, inhibition of mitochondrial energy metabolism (lowering of ATP) can change the mode of cell death from apoptosis to necrosis. Similarly, inhibition of caspase activity might change apoptosis to necrosis or autophagic cell death, whereas activation of calpain-mediated cleavage of autophagy-regulated protein, Atg-5, switches the mode of cell death from autophagy to apoptosis. Interestingly, although necroptosis, necrosis and secondary necrosis following

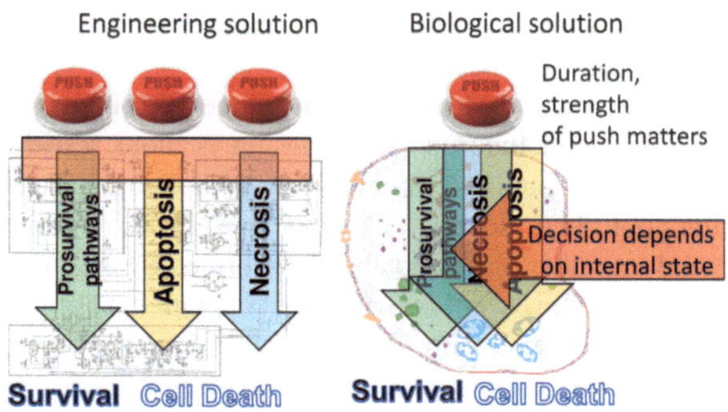

Fig. 6.1 Engineer vs. biological view of a cell decision device. Engineering solutions are characterised by clear separation of modules, maximum predictability of the response. Biological solutions are composed of overlapping modules, a response to a perturbation that is probabilistic and dependent on the internal system state

apoptosis, represents different modes of cell death, all of them might eventually converge on similar cellular disintegration features, albeit with different kinetics. It seems that the point of no return in various cell death modalities is also similar and associated with mitochondria. Expression of anti-apoptotic members of Bcl-2 family proteins rescues cells that undergo apoptosis, necrosis/necroptosis and autophagy. Thus, learning more about the molecular mechanisms regulating various cell death modalities and their crosstalk is important, since they play a critical role in multiple biological processes.

In recent articles, there appeared a number of reports in which analogies and differences between engineering and biological implementations of specific functions are discussed (Lazebnik 2002). If we were to study cell fate decision between survival, necrosis and apoptosis from the engineer's prospective, we would be asked to build a device capable of performing three mutually exclusive functions. The most probable design would be to associate to each output, one single corresponding input, as depicted in Fig. 6.1 (*left panel*). This device is such that it contains three modules performing three functions, each module being activated by a separate button. To improve the response, some further analyses of the state of the buttons would be needed: for example we would need to determine what happens if a button is pressed when another one is already pressed.

By contrast, the design that is used by nature is quite different (Fig. 6.1, *right panel*). It is known that the three modules are connected by extensive pathway crosstalks, and therefore with a high overlap. There is no evident separation between them as it would be the case for the engineer's modules. Therefore, the definition of one module or pathway and the decoupling of one module dynamics become difficult tasks. Moreover, nature may use only one button (receptor) to activate the three modules and the decision will then depend on the strength with which the button is pressed, the duration and the presence or not of some of the components of the pathways at the exact moment when the button is pressed. As a result, the response is stochastic and characterised by a non-zero probability of having several resulting phenotypes. If many buttons were to be used, then the output would be even less predictable.

To assess which design would be more advantageous, one has to formulate the optimality criteria, i.e. conclude on which design is more resistant to a random withdrawal of certain elements by systematically comparing both designs. Another possible criterion could be to identify which design is more probable to appear in the process of random tinkering. This can be performed by assuming that a typical mechanism of evolution consists of two recurrent operations: duplication of already existing and working elements with further partial specialisation and partial conservation of a common function; and an ad hoc re-use of already existing settings for a new and completely different purpose. This evolutionary process leads to the property of *degeneracy of the genetic networks* and to the phenomena of *network buffering* and *distributed robustness* (Whitacre and Bender 2010). With these considerations, several questions emerge, for example: why does cytochrome C play a role both in the respiratory chain of mitochondria and in the apoptosis signaling pathway after mitochondria membrane permeabilization has occurred? Is this double role an advantage or is it a mere random consequence of the ad hoc use of cytochrome C? How does this decision, when fixed, constrain further the evolution of the machinery, etc.?

The analogy between the two views brings into relief some unexpected questions. It mainly proves that much more should be understood in order to define the correct optimality criteria used by evolution and, hence, to separate the role of chance from the role of design and selection in biological evolution settings as compared to engineered devices.

6.3 Mathematical Modeling

Cell fate decision has raised interest in many areas, from immunology to haematopoiesis, to cell cycle events, or various signaling networks such as growth factor pathways, insulin pathway and apoptotic pathways. Recently, systems biology provided a number of concrete implementations of the Waddington's cell fate decision paradigm, through mathematical modeling of various biological systems. In this chapter we will focus on modeling cell fate decisions between life and death in mammalian cells.

There are several ways to study cell fate decision, from purely phenomenological approaches (Aguda and Algar 2003), to very detailed descriptions (Helikar et al. 2008). Here, we propose to approach the cell fate problem from a prospective of biochemistry and genetics by constructing a genetic network recapitulating, in a simplified form, our knowledge of the molecular processes leading to death or survival (Fig. 6.2). Based on a thorough literature search, we built an influence network that describes how the master regulators of cell decision, i.e. genes, complexes or processes, relate to each other in a global context. The two possible relations of the influence network are activation and inhibition. The master

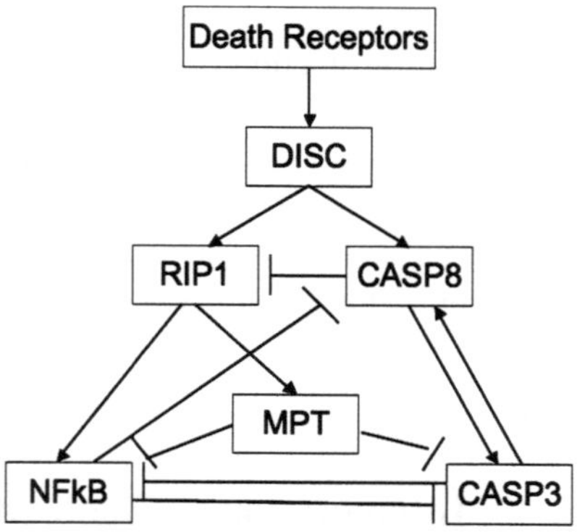

Fig. 6.2 Simple view of the cell fate decision model. From a single input, death receptor activation, three pathways are proposed: survival pathway with NFkB activation, necrosis with MPT occurring and apoptosis with CASP3 cleavage

regulators were chosen based on the importance of the role they play in the three pathways that we decide to describe in response to the engagement of death receptors: apoptotic and necrotic pathways for cell death and NFkB activation pathway for cell survival.

To be able to convert the genetic network into an analytical tool, we translate it into a mathematical language that will describe the possible dynamical behaviours in specific conditions. This description allows to determine the set of possible system solutions or outputs (understood as molecular processes such as activation of a protein or shutting down expression) and the sequence of events that leads to these solutions.

The general strategy of building and validating the model is the following:

1. Construction of the generic graph that describes the different processes of cell survival and cell death in normal conditions. The resulting genetic network recapitulates published knowledge on biochemistry.
2. Verification of the coherence of the network by crossing it with genetic results (mutants and drug treatments).
3. Formulation of predictions on the network topology, on mutant phenotypes, on alterations of pathways.
4. Experimental validation.

The mathematical formalism we chose to describe cell fate decisions in response to death signals is the Boolean formalism. Boolean modeling is useful when one wants to verify that the recapitulating network is coherent with published data. Since it is built from uncorrelated and disseminated articles, there is no guarantee that when the pieces of the puzzle are assembled and integrated into a unique and global picture, it accounts for the different expected phenotypes in all cell conditions. In other words, modeling is one way to prove that the construction of the network that is supposed to reproduce the cell behaviour of a disease or in altered signaling pathways is somehow based on coherent facts. Understanding the topology of the pathways that are altered in normal and abnormal conditions might shed some light on ways to counteract them. Boolean modeling is an abstraction of reality, and of continuous and more refined methods such as ordinary differential equations (ODEs) that would require more quantitative data (the concentration of proteins as a function of time, the reaction rates, and the speed of events). Since quantitative knowledge is hard to find and to include in the model, the description needs to remain as qualitative as possible. Cell fate decisions can be represented as a dynamic system having several qualitatively different states. Having this in mind and also because the quantitative parameters of such a system are rarely available, we assume that the process of cell fate decisions can be appropriately treated by Boolean formalism.

6.4 Introduction to Boolean Modeling of Cell Decisions

Boolean networks are primarily described by *regulatory* or *influence networks*. They consist of signed and directed graphs that are composed of nodes and arrows. The nodes, or variables, correspond to genes, proteins, complexes or processes depending on the information collected and summarised in the network. These variables can only take two values in a Boolean framework: 0 if the species represented by the node is absent or inactive, or 1 if it is present or active. Arrows correspond to interactions and provide two types of information: they can be activating (+) or inhibiting (−) arrows. A node is updated according to the state of its input nodes, the sign of the incoming arrows and the logical rule that connect the inputs. The most widely used logical connectors found in the published models are AND for which node A and node B need to be present for node C to be activated; OR for which node A or node B need to be present for node C to be activated; NOT for which node A needs to be absent for node C to be activated (Fig. 6.3).

The choice of the logical rules is, in the best cases, determined by biological knowledge. The dimension of the network is dependent on the amount of information that is gathered and, of course, on the biological question and the amount of details it requires to answer. Some models (Helikar et al. 2008; Schlatter et al. 2009) are able to explain decision making with great details and a high number of variables. Not only they are able to account for cell decision making in different cellular conditions, but they are also validated on experimental data. For these models, the logical rules for each node are deduced from experimental facts.

The added value of Boolean modeling lies in the fact that a simple formalism with limited amount of information can already explain—or start to explain—some complex and misunderstood observations. For instance, it can predict the effect or the necessity of some drugs in cancerous scenarios (Layek et al. 2011). Moreover, it is one way to show that the topology of a network can provide some insight into the regulation of particular biological processes.

Boolean modeling is a simple formalism for animating an influence diagram. To achieve this, the state transition graph associated to the Boolean model (a diagram of all the possible states and their sequence) is generated. On this graph, each node represents a state of the system, which in this case can be encoded by a sequence of 0's and 1's. The states are connected by a directed edge if a

AND	OR	NOT
A => C <= B	A => C <= B	A =| C
C = A & B	C = A | B	C = !A

Fig. 6.3 Three simple logical connectors: AND (also represented by and), OR (also represented by |), NOT (also represented by !)

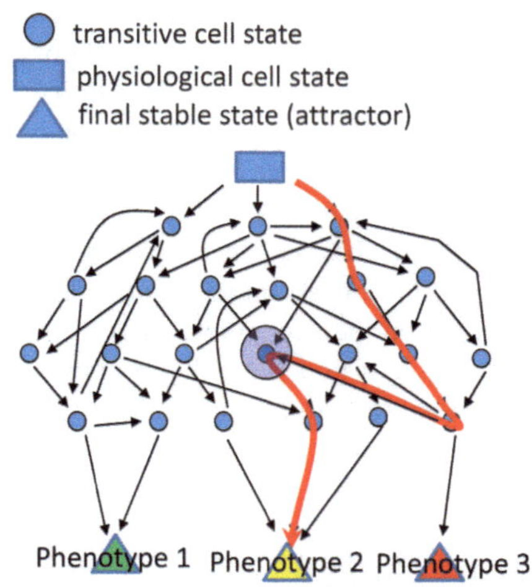

Fig. 6.4 State transition graph in a Boolean model. One trajectory from a physiological cell state to the Phenotype 2 is shown. Along this trajectory, a "point of no return" is *highlighted*. At this point, the choice of the output is determined

transition between two states is allowed, and according to any of the logical update rules predefined by the model. In principle, the state transition graph can be built independently and without the biological diagram; however, this would require a tremendous amount of empirical knowledge, which is not available. Hence, the biological diagram with associated Boolean rules is used as a compact representation and a tool to generate the state transition graph. More detailed instructions for this procedure can be found in Chaouiya et al. (2006).

The state transition graph recapitulates all the possible trajectories of the system's dynamics. It provides all the possible states of the system in a discrete phase space (Fig. 6.4). It can be compared to the epigenetic landscape introduced by Waddington. From an initial condition or physiological cell state, a certain path is followed, decisions are made at some bifurcating points and the ending point corresponds to the cell fate chosen by the cell. These ending points can also be called *sink* nodes. Any allowed transition from this node leads to itself. It is an attractor, a solution or a stable state solution of the model. Note that the state transition graph is assumed to be rather sparse compared to the fully connected graph.

The state transition graph can be seen as a tool to interrogate the model. For instance, one could find the probability to reach each stable state starting from a particular state. To do so, the state transition graph is converted into a Markovian process of random walk on a graph and analysed by classical techniques (Feller 1968).

The final probability is associated with a probability of observing a particular phenotype in an experiment. In this way, the Boolean model can take into account the stochastic nature of cellular decisions and be validated experimentally. However, due to the lack of quantitative information, the probabilities associated to each transition are considered to be equiprobable. One biological interpretation of these probabilities could be that for a generic cell, all possible system trajectories are explored with the same probability. For a specific cell type, these transition probabilities would need to be weighted according to the importance—or even the presence—of a transition. Of course, it can be argued that in any particular concrete cell, no event occurs with an equal probability and that the generic cell is not representative of anything real observed in an experiment. Having in mind this difficulty, direct interpretation of absolute values of probabilities should be avoided, concentrating rather on their relative changes in response to some system modifications such as removing a node or fixing a node's activity. The "generic" cell model is already capable of reproducing a number of known experimental facts.

6.5 Boolean Modeling of Cell Death

Our special interest here is Boolean modeling of cell death. Boolean formalism has been used in the context of cell death at different levels (Fig. 6.5): in death receptor-induced apoptosis (Mai and Liu 2009) and more specifically in cell response to death receptor engagement (Calzone et al. 2010; Tournier and Chaves 2009; Philippi et al. 2009); in the mechanisms observed in both intrinsic and extrinsic

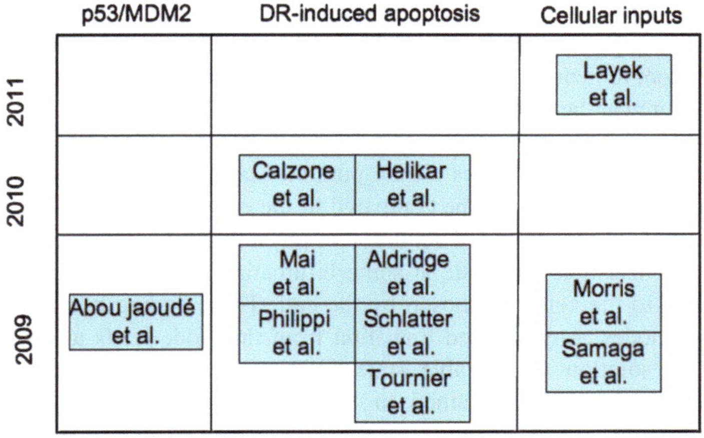

Fig. 6.5 Some examples of published Boolean models of cell death

apoptosis (Schlatter et al. 2009) and in the interaction between MDM2 and p53 (Abou-Jaoude et al. 2009).

Most of the models of cell death are built as a response to cell receptor activation. Some concentrate on death receptors such as Fas receptors (Philippi et al. 2009), TNF receptors (Tournier and Chaves 2009; Schlatter et al. 2011) or both (Calzone et al. 2010); others on growth factors (Layek et al. 2011; Morris et al. 2010) or combinations of both growth factors and death receptors (Mai and Liu 2009); on the synergy between diverse input signals: RTK, GPCR and integrins (Helikar et al. 2008) or as many as eight different inputs including insulin, glucagon, death receptors, interleukins, SMAC mimetics and UV irradiation (Schlatter et al. 2009). Because the p53/MDM2 module is governed by a negative feedback loop and that negative feedback loops are not as easy to interpret in Boolean formalisms, there are few models that describe the activation of the apoptotic pathway via the p53 pathway (Abou-Jaoude et al. 2009).

According to the mathematical model and the consequent degree of details, some genes are chosen as readouts of the activated pathways. They may not mean that the phenotype is observed in the cell, though, only that the activation of the corresponding pathway is possible. What happens downstream of these readouts is often not described and the fact that these genes are activated does not guarantee that a fate will be realised. That way, for instance, some genes, alone or in complexes, or some processes, are selected to represent outputs: Akt, CASP3, AIF or p53 may account for apoptosis; BCL-2, IAP or NFkB for survival; mitochondrial permeability transition (MPT) for necrosis; gene transcription for cell cycle activation; etc.

6.6 Cell Fate Decision Model

We have proposed a molecular mechanism that explains cell fate decision between life and death in response to death receptor engagement (Calzone et al. 2010).

The only inputs of the model are the activation of the two death receptors: TNFR and Fas. The different outputs or phenotypes considered in this model are the following (1) the resting state referred to as "naïve" for which the cell is not under any life-threatening stress conditions and not receiving death signals; (2) survival which is associated here to active resistance to stressful conditions by activation of the NFkB pathway (Karin 2006); (3) death by necrosis which is triggered by a death receptor-induced pathway and (4) apoptosis.

The pathways depicted here (Fig. 6.6) are the result of a tremendous simplification, the purpose of this study being the understanding of the cell decisions in response to a single input rather than the molecular details of the different pathways.

NFkB transcription factor is activated after the degradation of its inhibitor, IkB (not explicitly shown in the diagram), with the help of a specific kinase called IKK. IKK is itself activated after a protein, RIP1, in its ubiquitinated form (RIP1ub on the diagram). When NFkB is activated and transported into the nucleus, it can mediate

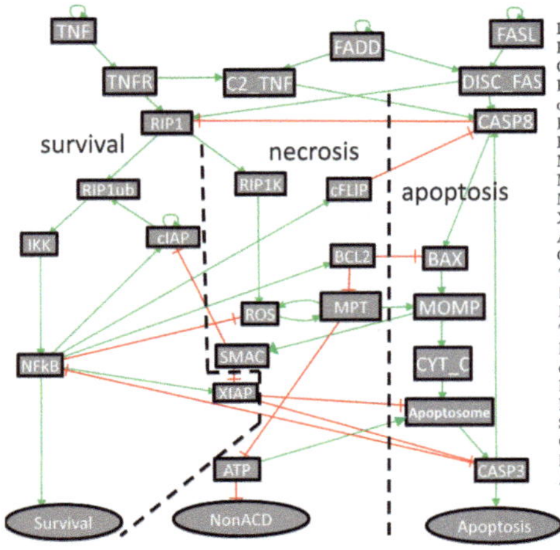

DISC_TNF' = TNFR AND FADD
RIP1' = (TNFR OR DISC_FAS) AND (NOT CASP8)
CASP8' = (DISC_TNF OR DISC_FAS OR CASP3) AND (NOT cFlip)
RIPub' = RIP1 AND cIAP
cIAP' = (NFkB OR cIAP) AND (NOT SMAC)
BAX' = CASP8 AND (NOT BCL2)
ROS' = (RIP1K OR MPT) AND (NOT NFkB)
MPT' = ROS AND (NOT BCL2)
MOMP' = BAX OR MPT
NFkB' = IKK AND (NOT CASP3)
XIAP' = NFkB AND (NOT SMAC)
Apoptosome' = CYT_C AND ATP AND (NOT XIAP)
CASP3' = Apoptosome AND (NOT XIAP)
TNF' = TNF
FADD' = FADD
FAS' = FAS
TNFR' = TNF
RIP1K' = RIP1
cFlip' = NFkB
IKK' = RIP1ub
BCL2' = NFkB
SMAC' = MOMP
CYT_C' = MOMP
DISC_FAS' = FAS AND FADD
ATP' = NOT MPT

Fig. 6.6 Boolean model of cell fate decisions between survival, apoptosis and necrosis. *Left panel*: Influence network of the molecular interactions involved in cell fate decisions. The diagram is roughly divided by *dashed lines* into three modules corresponding to three mechanisms of cell fate decisions. *Right panel*: the table of logical rules defining the discrete mathematical model is provided. To each node corresponds a logical rule

the transcription of a plethora of genes, among which are cFLIP, Bcl-2, ROS scavenging enzyme coding genes, members of the IAP family cIAP1, cIAP2 (together referred to as cIAPs) and XIAP.

There exists a big overlap between the necrotic and apoptotic pathways. The line separating them in Fig. 6.6 is not clearly defined. After the cell receives death receptor signals, the apoptotic pathway is activated by the formation of the DISC (either TNF- or Fas-death inducing signaling complex), leading to the cleavage of CASP8. When active, CASP8 can initiate the disruption of the mitochondria outer membrane with the help of a family of BAX proteins. As a result, many mitochondrial components are released into the cytosol. Among them, cytochrome C (CYT C) is one of the most important components for inducing apoptosis. CYT C is a member of the *Apoptosome* complex that leads to the activation of the executioner caspases such as CASP3. A series of event after CASP3 causes the destruction of intracellular structures including membranes and DNA. Another important ingredient of the mitochondria inter-membrane space is the second mitochondria-derived activator of caspase (SMAC), for its role in inhibiting survival.

Necrosis is tightly linked to reactive oxygen species (ROS), which, when present in high concentrations, can severely damage mitochondria and cause MPT. MPT in turn causes an additional increase in ROS in the cytosol. In our diagram, ROS is also controlled by RIP1K, the kinase activity of RIP1, but the level of ROS is not solely controlled by RIP1. A high increase in the concentration of ROS can be a consequence of many reasons such as toxic conditions, for instance. Thus, in our context, necrosis can be initiated in a programmed fashion. Disrupting mitochondria in a cell leads to severe deficit of ATP, the main biochemical currency, with lethal consequences. A drop of ATP is understood as a necrotic phenotype.

It is important to underline that the three pathways are connected by multiple crosstalks (Fig. 6.6) at all levels of the pathways (upstream and downstream). These crosstalks raise a question about the definition of each of the pathways, since it is difficult to decompose the network back into independent modules without an a priori definition of pathway borders.

The fully assembled graph is an influence network composed of nodes and arcs. The nodes represent a protein (TNF, FADD, FASL, TNFR, CASP8, cIAP, cFLIP, Bcl-2, BAX, IKK, NFkB, CYT C, SMAC, XIAP, CASP3), a state of protein (RIP1ub, RIP1K), a small molecule (ROS, ATP), a molecular complex (apoptosome, C2 TNF, DISC FAS), a group of molecular substances (Bax, Bcl-2), a molecular process (MPT, mitochondrial outer membrane permeabilization (MOMP)) or a phenotype (survival, apoptosis, non-apoptotic cell death, nonACD). The arcs represent influences of a node onto another node, either positive (arrowed edge) or negative (headed edge).

As indicated in the previous section, the influence network is translated into a dynamical model using Boolean formalism. To each node of the diagram, a logical rule is associated (Fig. 6.6, *right panel*).

The scope of this model is to identify the main players and the possible routes that lead to different phenotypes in response to the same signal. Death receptor activation such as Fas or TNFR can lead to cell death (through necrosis and apoptosis) or survival (by activation of NFkB) according to cell condition and cell type. Moreover, the model emphasises the necessity of mutual inhibition in these pathways to ensure a proper and unique cell response. Tournier and Chaves (2009) have shown a similar model between NFkB activation and apoptosis.

6.7 In Silico Experiments

The simulation of the model is based on the properties of random walk on the state transition graphs as mentioned previously. We computed the probability of reaching one or the other phenotype, considered as a stable steady state of the system, for wild-type conditions and for perturbed conditions (mutants or drug treatments). Mutant models include several modifications: gene deletions that are simulated in the Boolean model by setting the corresponding node or variable to 0, i.e. insensitive to its inputs or external stresses; overexpressions that are simulated in a similar fashion by setting the node to 1; drug treatments by setting the targets of the drug to the appropriate value on the concerned variables. The result of these simulations is provided in Fig. 6.7.

In the paper by Calzone and colleagues (2010), this table was systematically compared to published experimental observations of the cell death phenotype modifications observed in various mutant systems, such as cell cultures and mice. The model was able to qualitatively recapitulate all of them. The model provides a

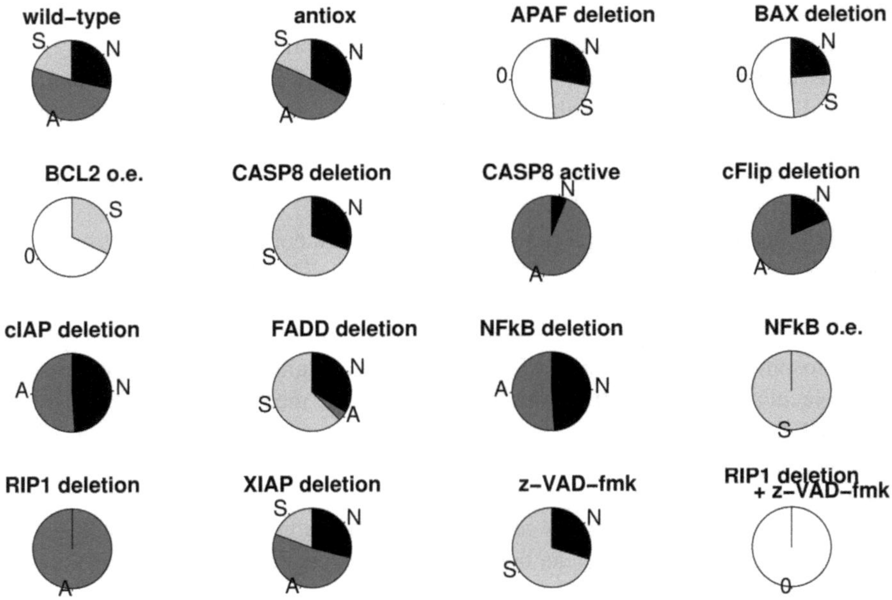

Fig. 6.7 Model probabilities of having a particular phenotype starting from the physiological condition

way to test the effect of more than one alteration. We could answer questions of the type: how many and which mutations are needed to re-establish a lost phenotype.

Various "mutant" modifications of the dynamical system are tested here in response to TNF activation. "A" denotes Apoptosis, "N" denotes Necrosis and "S" denotes Survival, "0" denotes naive state, "o.e." stands for overexpression of a protein, "antiox" corresponds to blunting the capacity of NFkB to prevent ROS formation, "z-VAD fmk" simulates the effect of caspase inhibitor z-VAD-fmk.

For all simulations presented in Fig. 6.7, TNF is set to 1 and therefore considered to be always ON along the trajectories. It is implied here that the death receptor is engaged for a sufficiently long time with a dose high enough to trigger a cell response. All trajectories are generated from a "physiological" condition where all variables are set to 0 except for TNF, cIAP and ATP.

We tested further in silico experiments by considering a temporary pulse of TNF rather than a sustained signal. Since time is discrete in our framework, each "time" or "event" corresponds to one step in the transition graph. The trajectories for all wild-type and perturbed systems are computed from the physiological condition. At a given time or after a certain number of events, the value of TNF is forced to 0. The probabilities to reach a phenotype are then calculated with the new initial condition that corresponds to the one at which TNF was switched to 0. The average probabilities are represented for the wild-type and the perturbed models (Fig. 6.8).

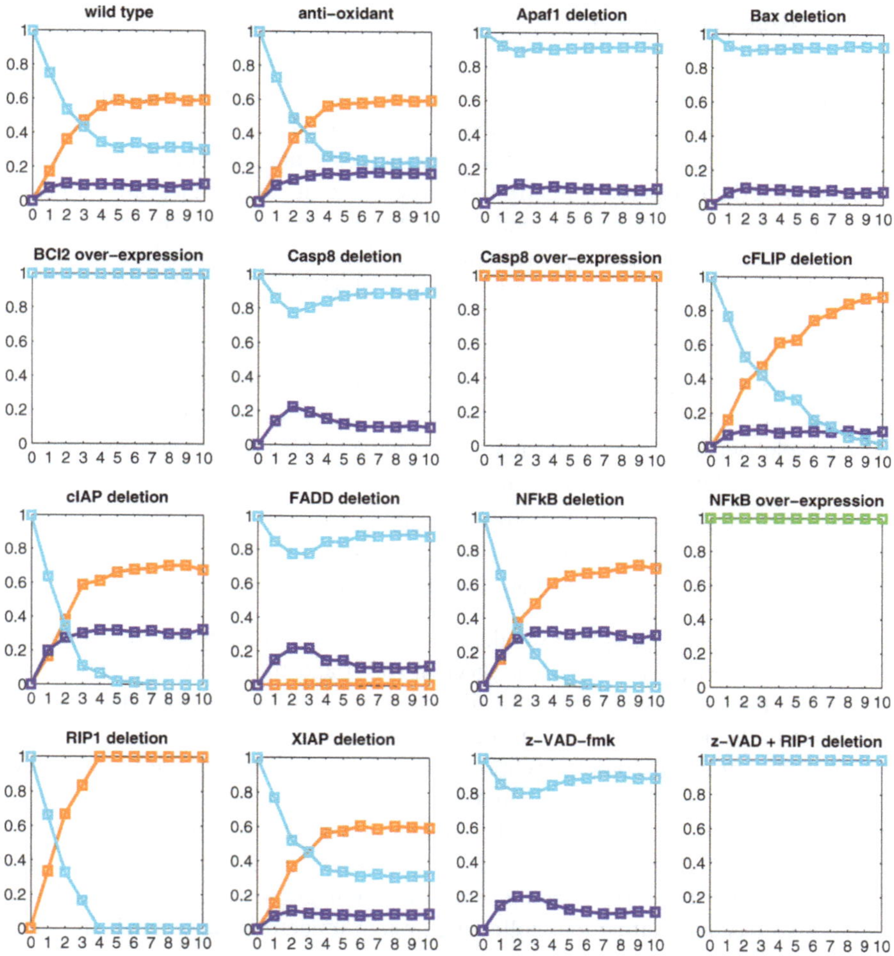

Fig. 6.8 Ligand removal experiments. The x-axis represents the (discrete) duration of the TNF pulse. At each discrete time point along the x-axis, the TNF signal is turned off. The different curves represent the average probabilities to reach the different attractors after the pulse (the number of trajectories $N = 2{,}000$). Curves are coloured in *blue* for naïve state, *green* for NFκB survival, *orange* for apoptosis and *purple* for necrosis

This experiment allows to conclude on the number of steps needed to achieve cell decision upon TNF treatment. It provides a measurable way to assess the appearance or disappearance of certain phenotypes upon TNF induction. The first time point corresponds to TNF OFF at all time. The last time point of Fig. 6.8 corresponds to the TNF sustained case (TNF ON all the time) presented in Fig. 6.7. For most cases, the decision is made very early, at step 4. It can be concluded that after a certain point, the cell has committed to its fate, even if the signal is removed.

6.8 Model Reduction

To complete our study, we reasoned on the simplest model of cell fate that can be deduced from the model described above. We selected the three readouts for our phenotypes: NFkB for survival, MPT for necrosis and CASP3 for apoptosis. We then listed all possible paths from one species to the other in a semi-automatic manner (Zinovyev et al. 2008), and we deduced a three-node influence diagram where the sign of each arc represents the overall sign of all the paths linking two nodes. In some ambiguous cases (e.g. influence of MPT on CASP3 or of NFKB on MPT), the decision on the sign of the influence is based on the Boolean rules and not on the paths only. In the case of mutations eliminating all the negative influences, however, a positive arrow must be considered. In Fig. 6.9, the complete network along with the state transition graph is shown (upper left and right). The simple network is symmetrical. Each node auto-activates and inhibits the other two. The logical rules are written as follows:

```
CASP3 = NOT MPT AND NOT NFkB AND CASP3
MPT = NOT CASP3 AND NOT NFkB AND MPT
NFkB = NOT CASP3 AND NOT MPT AND NFkB
```

The state transition graph shows the existence of four phenotypes corresponding to apoptosis (100), necrosis (010), survival (001) and the passive naïve survival (000). The three phenotypes are mutually exclusive. This is coherent with what was observed in the analysis of the complete model.

From this minimal complexity model, one can conclude that the initial model contains three mutually inhibiting and self-activating "modules" insuring stability and separability of the phenotypes. It is not easy to dissect and clearly separate the modules at the level of genetic network. However, one can demonstrate that the structure of the corresponding state transition graph can be separated into three large basins of attraction, collecting system trajectories and canalising them to a particular phenotype.

In the language of engineers, the motif shown in Fig. 6.9 can be called a "three-stable trigger". Bistable triggers were shown to be a typical motif in other molecular mechanisms, regulating, for example the cell cycle (Santos and Ferrell 2008).

6.9 Concluding Remarks

Systems biology takes an interdisciplinary approach to the systematic study of complex interactions in biological systems. This approach seeks to decipher the emergent behaviours of complex systems rather than focusing only on their constituent properties. As one example, systems biology approaches help to understand the initiation, progression and execution of various cell death modalities. Mathematical models, as an important part of systems biology, provide a tool to investigate all possible scenarios in the cell and propose networks or mechanisms

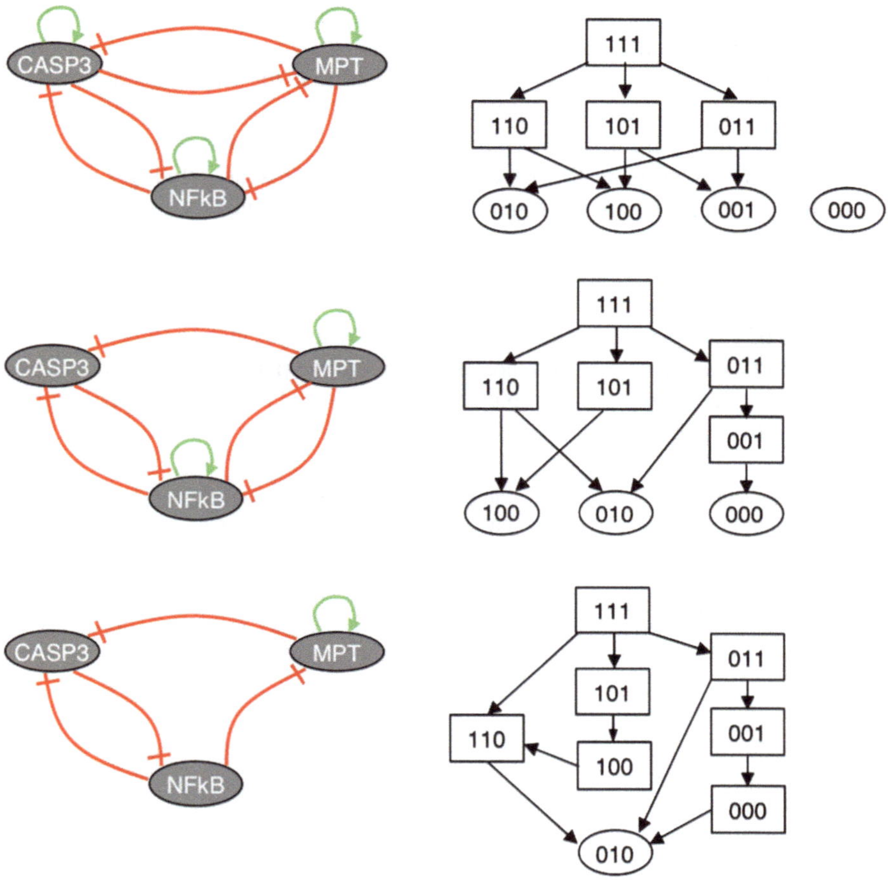

Fig. 6.9 A minimal model of cell fate decision process. *Upper row*: minimal conceptual model of cell fate decision machinery (*left*) and its state transition graph (*right*). *Middle row*: a modification of the minimal model, obtained by removing CASP8 from the biological diagram. *Bottom row*: a modification of the minimal model, obtained by removing both CASP8 and cIAP from the diagram (an example of synthetic interaction)

recapitulating experimental observations. They permit to translate biological knowledge into a mathematical language. One of the main purposes of these models is to shed light on some contradictions and paradoxes described in the literature, to summarise, as much as possible, knowledge about a specific process and to give some insight on obscure phenomena. Indeed, any kind of hypotheses can be tested in silico before performing experiments in the laboratories, when one is confident enough with the model. For a model to be used as a tool for biologists, it must reproduce existing data, from wild-type conditions to mutant phenotypes and drug treatment outputs. This requires tight interactions between theoreticians and experimentalists, and short feedback loops from model results to experimental assessment, and model revision if needed. Only then can the modeller or the biologist trust the model and formulate predictions.

Since the main goal of systems biology is to find the most rational route, a unified "language" (model) should be used by bioinformaticians, which should include all available information essential for understanding the proper functioning of the biological system. In our case, the model aims at defining the proper functioning of cell death machinery. As mentioned above, dysfunction of cell death programme leads to various diseases associated with too much or too little cell death. Therefore, systemic understanding of cell death pathways along with development of new methods of data analysis will result in refining the molecular diagnosis of various types of disease, associated with dysfunction of cell death, of optimising the calculation of prognostic and predictive parameters, of guiding new strategies for the amelioration of existing treatments and the identification of novel targets for therapeutic modulation of cell death pathways.

Acknowledgement The work in the authors' laboratories was supported by grants from the EC FP7 (Apo-Sys) programme. BZ is also supported by the Swedish Research Council, the Swedish and the Stockholm Cancer Societies, the Swedish Childhood Cancer Foundation and the EC FP-6 (Chemores) programme. LC and AZ are members of the team "Systems Biology of Cancer," Equipe labellisée par la Ligue Nationale Contre le Cancer. We also acknowledge support from Agence Nationale de la Recherche (project ANR-08-SYSC-003 CALAMAR) and from the Projet Incitatif Collaboratif "Bioinformatics and Biostatistics of Cancer" at Institut Curie.

References

Abou-Jaoude W, Ouattara DA, Kaufman M (2009) From structure to dynamics: frequency tuning in the p53-Mdm2 network I logical approach. J Theor Biol 258(4):561–577

Aguda BD, Algar CK (2003) A structural analysis of the qualitative networks regulating the cell cycle and apoptosis. Cell Cycle 2(6):538–544. doi:550 [pii]

Balazsi G, van Oudenaarden A, Collins JJ (2011) Cellular decision making and biological noise: from microbes to mammals. Cell 144(6):910–925

Calzone L, Tournier L, Fourquet S, Thieffry D, Zhivotovsky B, Barillot E, Zinovyev A (2010) Mathematical modelling of cell-fate decision in response to death receptor engagement. PLoS Comput Biol 6(3):e1000702

Chaouiya C, de Jong H, Thieffry D (2006) Dynamical modeling of biological regulatory networks. Biosystems 84(2):77–80

Degterev A, Hitomi J, Germscheid M, Ch'en IL, Korkina O, Teng X, Abbott D, Cuny GD, Yuan C, Wagner G, Hedrick SM, Gerber SA, Lugovskoy A, Yuan J (2008) Identification of RIP1 kinase as a specific cellular target of necrostatins. Nat Chem Biol 4(5):313–321

Feller W (1968) An introduction to probability theory and its applications, vol 1. Wiley, New York

Golstein P, Kroemer G (2007) Cell death by necrosis: towards a molecular definition. Trends Biochem Sci 32(1):37–43

Helikar T, Konvalina J, Heidel J, Rogers JA (2008) Emergent decision-making in biological signal transduction networks. Proc Natl Acad Sci USA 105(6):1913–1918

Karin M (2006) Nuclear factor-kappaB in cancer development and progression. Nature 441 (7092):431–436

Kerr JF, Wyllie AH, Currie AR (1972) Apoptosis: a basic biological phenomenon with wide-ranging implications in tissue kinetics. Br J Cancer 26(4):239–257

Kroemer G, Galluzzi L, Vandenabeele P, Abrams J, Alnemri ES, Baehrecke EH, Blagosklonny MV, El-Deiry WS, Golstein P, Green DR, Hengartner M, Knight RA, Kumar S, Lipton SA, Malorni W, Nunez G, Peter ME, Tschopp J, Yuan J, Piacentini M, Zhivotovsky B, Melino G (2009) Classification of cell death: recommendations of the Nomenclature Committee on Cell Death 2009. Cell Death Differ 16(1):3–11

Layek R, Datta A, Bittner M, Dougherty ER (2011) Cancer therapy design based on pathway logic. Bioinformatics 27(4):548–555

Lazebnik Y (2002) Can a biologist fix a radio?–Or, what I learned while studying apoptosis. Cancer Cell 2(3):179–182

Lockshin RA, Williams CM (1965) Programmed cell death–I cytology of degeneration in the intersegmental muscles of the Pernyi Silkmoth. J Insect Physiol 11:123–133

Mai Z, Liu H (2009) Boolean network-based analysis of the apoptosis network: irreversible apoptosis and stable surviving. J Theor Biol 259(4):760–769

Morris MK, Saez-Rodriguez J, Sorger PK, Lauffenburger DA (2010) Logic-based models for the analysis of cell signaling networks. Biochemistry 49(15):3216–3224

Philippi N, Walter D, Schlatter R, Ferreira K, Ederer M, Sawodny O, Timmer J, Borner C, Dandekar T (2009) Modeling system states in liver cells: survival, apoptosis and their modifications in response to viral infection. BMC Syst Biol 3:97

Santos SD, Ferrell JE (2008) Systems biology: on the cell cycle and its switches. Nature 454(7202):288–289

Schlatter R, Schmich K, Avalos Vizcarra I, Scheurich P, Sauter T, Borner C, Ederer M, Merfort I, Sawodny O (2009) ON/OFF and beyond–a boolean model of apoptosis. PLoS Comput Biol 5(12):e1000595

Schlatter R, Schmich K, Lutz A, Trefzger J, Sawodny O, Ederer M, Merfort I (2011) Modeling the TNFalpha-induced apoptosis pathway in hepatocytes. PLoS One 6(4):e18646

Tournier L, Chaves M (2009) Uncovering operational interactions in genetic networks using asynchronous Boolean dynamics. J Theor Biol 260(2):196–209

Waddington C (1957) The strategy of the genes. George Allen and Unwin, London

Whitacre JM, Bender A (2010) Networked buffering: a basic mechanism for distributed robustness in complex adaptive systems. Theor Biol Med Model 7:20

Wirawan E, Vande Walle L, Kersse K, Cornelis S, Claerhout S, Vanoverberghe I, Roelandt R, De Rycke R, Verspurten J, Declercq W, Agostinis P, Vanden Berghe T, Lippens S, Vandenabeele P (2010) Caspase-mediated cleavage of Beclin-1 inactivates Beclin-1-induced autophagy and enhances apoptosis by promoting the release of proapoptotic factors from mitochondria. Cell Death Dis 1(1):e18

Zinovyev A, Viara E, Calzone L, Barillot E (2008) BiNoM: a cytoscape plugin for manipulating and analyzing biological networks. Bioinformatics 24(6):876–877

Chapter 7
Modeling Single Cells in Systems Biology

Nicolai Fricker and Inna N. Lavrik

Abstract One of the most powerful methods to study the dynamic behavior of protein networks is a single-cell analysis. Introduction of fluorescent proteins provided phenomenal approach to follow living cells in the spatiotemporal manner. In this chapter we shall discuss major principles and tools used in single-cell analysis of apoptotic cells. To understand why single-cell analysis is so important, we shall compare advantages and disadvantages of single-cell versus bulk measurements. Furthermore, we shall discuss the models based on the live cell imaging and information that can be obtained with these models. In particular, we shall focus on the studies devoted to the dynamics of caspase activation and mitochondrial outer membrane permeabilization.

N. Fricker
Institute of Bioinformatics and Systems Biology, Helmholtz Zentrum München, Ingolstädter Landstraße 1, 85764 Neuherberg, Germany

Division of Immunogenetics, German Cancer Research Center (DKFZ), Heidelberg 69120, Germany

Bioquant, Heidelberg 69120, Germany

I.N. Lavrik (✉)
Department of Translational Inflammation Research, Institute of Experimental Internal Medicine, Otto von Guericke University, Leipziger Str. 44, 39120 Magdeburg, Germany
e-mail: inna.lavrik@med.ovgu.de

I.N. Lavrik (ed.), *Systems Biology of Apoptosis*, DOI 10.1007/978-1-4614-4009-3_7,

7.1 Introduction

Single cells of a genetically homogeneous cell population respond differently upon perturbation with a stimulating agent (Spencer et al. 2009). Stimulation of a cell population with death receptor (DR) ligand, for instance, will only induce death in a fraction of cells, whereas some cells will survive. There are a number of reasons for this nongenetic cell-to-cell variability. One of them is the phase within the cell cycle, which a particular cell has. A drug which induces DNA damage in S phase does not harm cells in G1 or G2 phase. Furthermore, even cells of a synchronized clonal population will show cell-to-cell variability. Sources of nongenetic cell-to-cell variability are cell size, cell density, stochastic fluctuations of biochemical reactions, and differences in the expression levels of proteins.

During the last years especially stochastic differences in protein expression have been investigated. The expression level of any protein is not identical among the cells of any cell population but rather forms a log-normal distribution as shown in Fig. 7.1 (Spencer et al. 2009). The expression level of a particular protein among cells of a clonal population can vary about 2.5-fold. This is not only true for cancer cells in culture but also for human tissue (Spencer et al. 2009). Where are these differences coming from? One of the sources of variations results from the differences at transcription level. In one cell there might be but few transcription initiating complexes, in some cells only one or two (Raj and van Oudenaarden 2008). Therefore, stochastic effects could lead to differences in gene transcription among single cells. Fluctuations in transcribed mRNA levels would result in differences in protein amounts. Naturally differences in protein amounts may give rise to differences in the response of the cells.

What might be the biological purpose of this nongenetic cell-to-cell variability? Sorger and Spencer suggest that this will give the human body a possibility to scale the response to a death stimulus (Spencer and Sorger 2011). For example, a certain amount of CD95L could induce death in a certain fraction of cells. If nongenetic cell-to-cell variability would not occur, either all or none of the cells would die. However, as cell-to-cell variability exists, the body can choose to induce cell death in, for example, 30% of all cells of a certain type. To induce apoptosis to a higher extent of the same population, the concentration of the death-inducing agent should be increased.

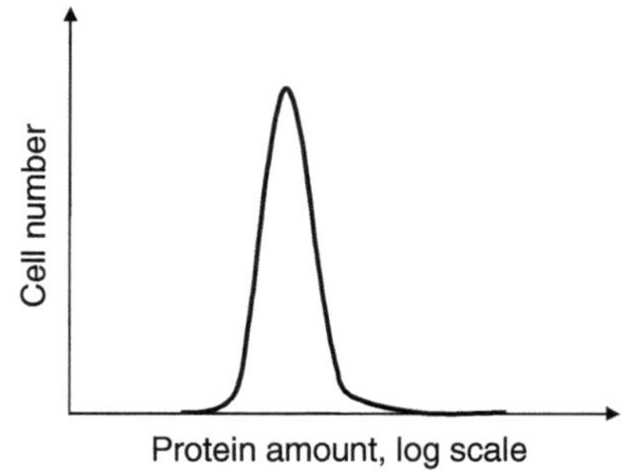

Fig. 7.1 Log-normal distribution of protein expression levels. The figure shows a scheme of the distribution of the expression level of a protein within a clonal population

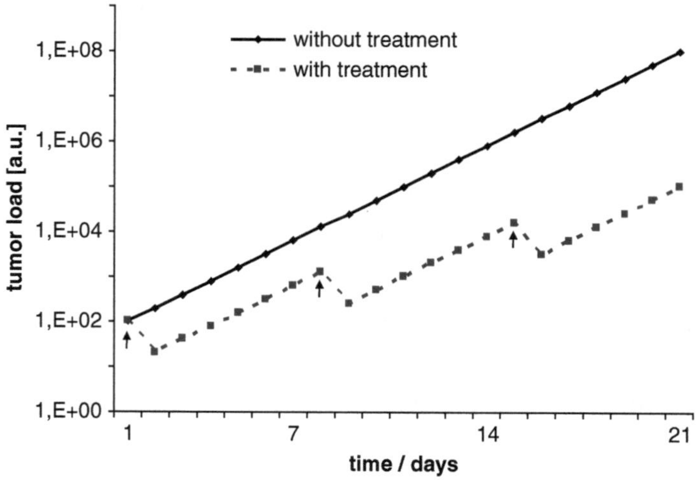

Fig. 7.2 Cell-to-cell variability contributes to tumor resistance. The *black curve* shows simulated tumor growth without treatment. The *gray dashed curve* displays tumor growth upon weekly treatment. *Arrows* indicate application of treatment. Cells in this model divide once per day. A single dose of treatment eliminates 90% of the tumor cells

While this interpretation is a pure speculation, the consequences of nongenetic cell-to-cell variability for treatment of cancer patients are of high importance. Upon treatment with a drug which should eliminate cancer cells, it is well documented that only a fraction of these cells would respond. A simple calculation shows that the assumption of only small difference in protein amounts, leading to the unresponsiveness of the marginal fraction of cells, might explain the absence of response to the treatment at the population level. Indeed, if one considers that upon treatment with a particular drug 10% of tumor cells would survive due to a nongenetic variability, then this particular drug would kill only 90% of tumor cells upon each application. Furthermore, taken that tumor cells divide once per day and the drug is applied once per week, this treatment could not control tumor growth in a patient—even in the absence of resistant stem cells or mutations! In the time interval during the application of treatment, more tumor cells will grow than will be killed by the drug (Fig. 7.2). This example demonstrates that to understand the population response, it is essential to follow the single cells. Therefore, it is of high importance to model the behavior of not only a certain tissue or cell type, but also of single cells.

7.2 Single-Cell and Bulk Measurement Techniques

To model protein networks two types of quantitative experimental data could be used: bulk measurements and single-cell analysis. In the following we will compare these two types of experimental data.

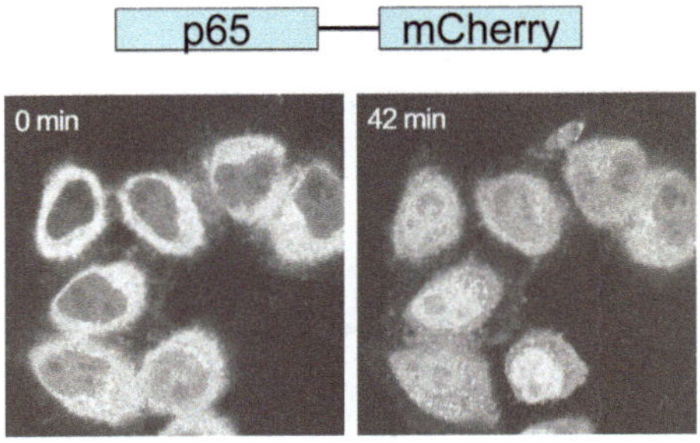

Fig. 7.3 Measuring NFκB localization with a p65-mCherry fusion protein. A scheme of the p65-mCherry fusion protein is displayed on *top*. *Bottom*: p65-mCherry is mainly found in the cytosol in non-stimulated cells. Forty-two minutes after CD95 stimulation with 1 μg/ml of agonistic anti-APO-1 antibody, p65-mCherry translocates into the nucleus. Modified from Neumann et al. (2010)

The most commonly used bulk measurement technique for protein amounts, their activity, and modifications is Western blotting (WB). It allows to determine the average protein amounts as well as cleavage, phosphorylation, or other protein modifications in a cell population (Schilling et al. 2005). WB is often combined with immunoprecipitations which allows to purify and analyze protein complexes. The principle is that the protein of interest can be detected with specific antibodies and further quantified using fluorescence or luminescence measurements. Furthermore, the number of proteins that could be detected simultaneously is unlimited if corresponding antibodies are available. The WB allows to follow the change in protein amounts, kinetics of protein modifications, and formation of protein complexes in the cell. However, WB is not applicable for single-cell measurements as the protein amounts produced by one cell are much too low to be detected.

Another method which can be considered as a bulk measurement is flow cytometry. In flow cytometry cells are marked or stained with fluorescent molecules or antibodies. This allows to determine the amount of particular protein in these cells. Several proteins may be measured simultaneously. In the field of cell death, a standard application of flow cytometry is determining the amount of dead or apoptotic cells in response to treatment. However, a disadvantage of this technique is that it is impossible to follow one single cell over a period of time. Flow cytometry rather takes a snapshot of a cell population at a certain time point. This disadvantage can be overcome with live imaging of a single cell.

Fluorescent tagging of proteins or probes has provided a remarkable insight into cellular processes. A protein can be fused to green fluorescent protein (GFP) or other fluorescent protein, which would yield information about the localization and amount of the fusion protein. This gives a unique way for tracking living cells, organelles, and even single molecules in the spatiotemporal manner. Following localization, fluorescent protein could be used as a functional readout for many cellular processes. For example, NFκB activation could be monitored by live cell imaging, e.g., NFκB nuclear translocation could be visualized using a p65-mCherry construct (Neumann et al. 2010) (Fig. 7.3).

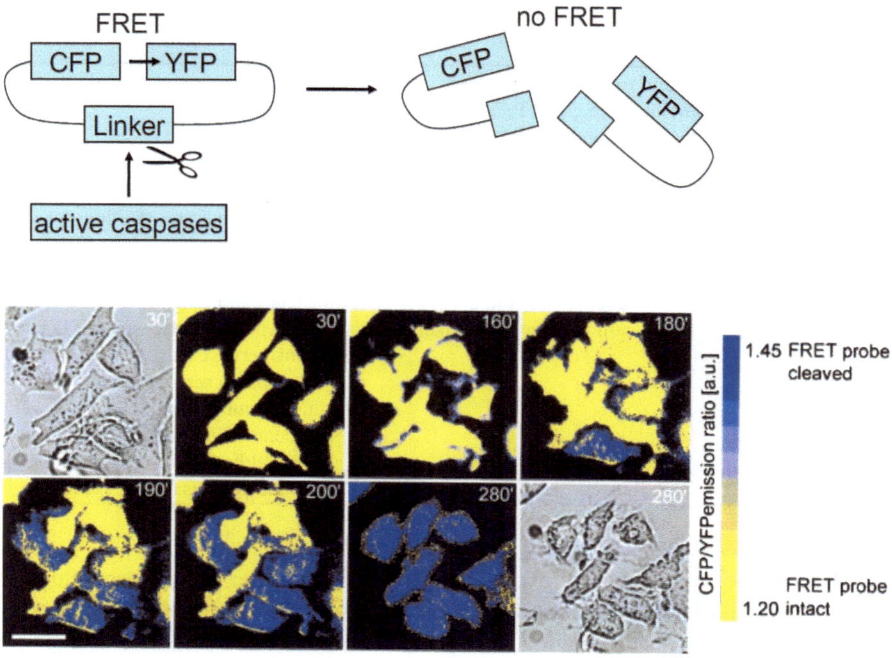

Fig. 7.4 Measuring caspase activity with FRET probes. *Top*: A scheme of a FRET caspase activity probe. It consists of two fluorophores linked by a peptide containing the cleavage sequence of a specific caspase. Upon cleavage of the linker by caspases, the fluorophores separate and the FRET signal is lost. *Bottom*: Loss of FRET signal of a caspase-3 activity probe as determined in HeLA cells upon stimulation with 3 μM staurosporine. Modified from Rehm et al. (2002)

Another method to visualize cellular events is the usage of activity probes, which typically comprise fluorescent proteins and could be modified by protein activity. In the field of apoptosis, activity probes are mostly constructed in a way such that they contain caspase cleavage site, which is cleaved upon caspase activation. Hence, these probes could be implemented to detect caspase activation and thereby apoptosis induction. One of the first caspase activity probes has been constructed by Rehm and coworkers (Rehm et al. 2002). It comprises two fluorophores, which could generate Foerster resonance energy transfer (FRET) signal. Two fluorophores are separated by the caspase cleavage site (Fig. 7.4). Without apoptosis induction the probe produces a constant FRET signal. Upon cleavage of the probe by caspases, the two fluorophores of the probe get separated and the FRET signal disappears (Fig. 7.4). In this way the decrease in FRET signal serves as readout for caspase activity.

Another way to determine caspase activity in single cells is to construct localization probes. The elegant localization probe was constructed by Joel Beaudouin (Fricker et al. 2010). It consists of a nuclear export signal linked to the cleavage sequence of a caspase fused to mCherry. In resting cells, the probe localizes to the

Table 7.1 Fluorescent probes that could be used to measure single cells

Type	Description	Example
Fusion proteins	A protein fused to a fluorescent molecule (e.g., GFP, YFP, CFP, mCherry, and others)	p65-mCherry (Neumann et al. 2010) (Fig. 7.3)
FRET probes	Two fluorophores are kept in proximity by a linker. Modification of the probe abolishes or enhances the FRET signal	Caspase-3 activity probe (Rehm et al. 2002). Cleavage of the linker by caspases separates the fluorophores, leading to the loss of the FRET signal (Fig. 7.4)
Localization probes	A target sequence fused to a fluorescent protein via a linker	A nuclear export signal fused to the linker peptide sequence IETD fused to mCherry. Cleavage of the linker by caspases separates the fluorescent protein from the target sequence. The fluorescent protein then freely diffuses and translocates into the nucleus (Fricker et al. 2010)

cytoplasm due to the nuclear export signal. Upon cleavage by caspases, nuclear export signal is cleaved off and mCherry translocates to the nucleus. Using fluorescent probes and live cell imaging, it is possible to follow the response of a single cell for hours or even for days. Table 7.1 summarizes different fluorescent probes that could be used in the field of apoptosis to measure single cells.

One might ask why WB is still used in quantitative biology and not exclusively live cell imaging. Fluorescent reporters have some disadvantages: The number of reporters that can be visualized simultaneously is, of course, limited. At some point the fluorescent spectra of the multiple fluorophores used will overlap. Furthermore, usage of a reporter requires manipulation of the respective cell. Introduction of the reporter or the reporter itself may alter cellular behavior. These limitations do not apply for WB. In addition, the formation of several cleavage products may be measured quite easily with WB. For instance, generation of the caspase cleavage products can be detected using WB. With fluorescent microscopy, there is no simple way to observe this. Therefore, it is always an optimal strategy to combine bulk measurements and single-cell experiments.

7.3 Relying Exclusively on Bulk Data Might Result in Wrong Interpretations

It is impossible to accurately model single-cell behavior when relying on bulk measurements exclusively. This can be shown with an example. In a bulk measurement, one could see a continuous and slow increase in protein concentration, for instance, of cleaved caspase-3 (Fig. 7.5a). Do the single cells display the same behavior? Not necessarily. Figure 7.5b shows two examples of responses of single

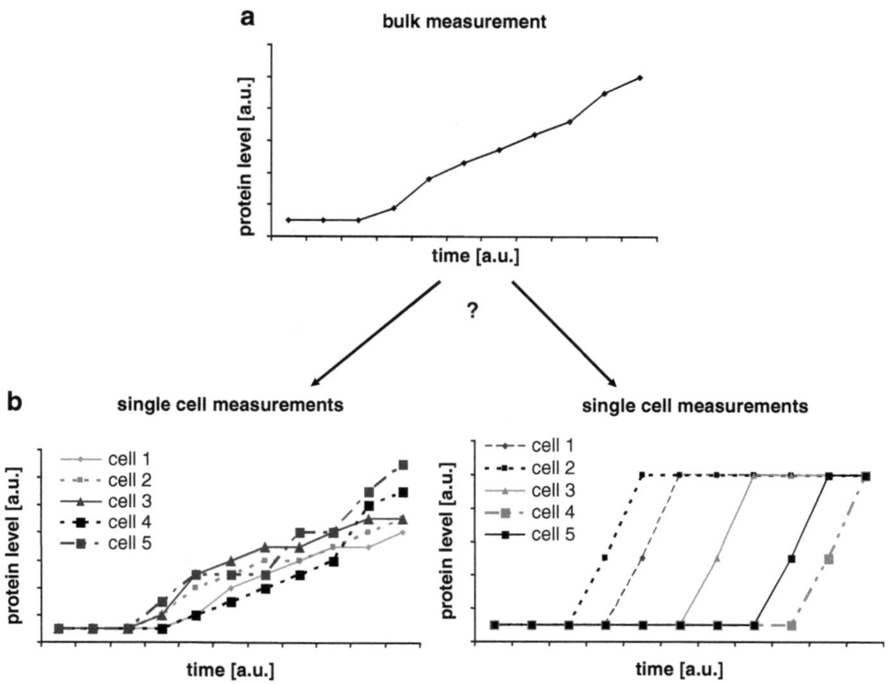

Fig. 7.5 Single-cell behavior cannot be recovered from bulk measurements. (**a**) Example of protein amounts determined in a bulk measurement. (**b**) Two examples of how single cells could behave resulting in the same bulk measurement data as shown on *top*

cells, which could result in the same bulk measurement. They look quite different. On the left-hand side of Fig. 7.5b, single cells display a slow and gradual increase in the amount of the protein. On the right-hand side, single cells show a variable lag time followed by a rapid response. Therefore one might draw wrong conclusions relying exclusively on WB.

Another example of events which are not feasible to analyze using bulk measurements are oscillations. For example, p53 oscillations were reported to occur upon DNA damage (Loewer et al. 2010). Figure 7.6 shows an example, which includes a bulk measurement and the corresponding single-cell data. In this case oscillations within single cells disappear in the bulk measurement. This is due to the fact that cells do not respond synchronously. Relying on bulk measurements, it is not possible to determine whether two events take place in the same or in different cells. Upon CD95 stimulation phosphorylation of IκBa and cleavage of procaspase-8 can be measured using WB (Fig. 7.7a). Does one cell induce apoptosis whereas another one decides to activate NFκB? Only with help of a fluorescent NFκB reporter our group could show that both pathways are indeed activated simultaneously in the same cell (Neumann et al. 2010) (Fig. 7.7b).

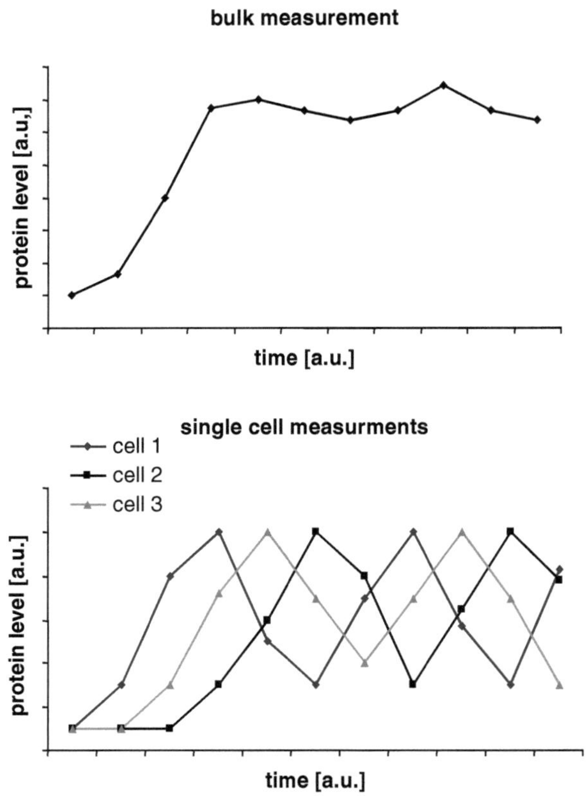

Fig. 7.6 Oscillations can be lost in bulk measurement. *Left*: Experimental data that could be obtained from a bulk measurement. There is no oscillation visible. *Right*: Single cells show oscillations. Building the mean value of the single-cell data will result in the bulk data as shown on the *top*

7.4 Modeling Single Cells

There are several considerations one has to take into account when modeling single cells in comparison to building models on bulk data. In the optimal case one knows the mean protein amounts of the main signaling molecules within a given population. These data can be directly put as initial protein concentrations into a model based on bulk measurement data. This rule does not apply to generation of models based on a single-cell analysis. One does not know the protein concentrations in a single cell. Furthermore, the protein amounts within the single cell might be 2.5 times higher or lower than the average amount within the population. Therefore, protein amounts of single cells have to be estimated. As a consequence single-cell modeling is more computationally demanding than population modeling. To fit ODE models to measurement data obtained from single cells, hundreds of

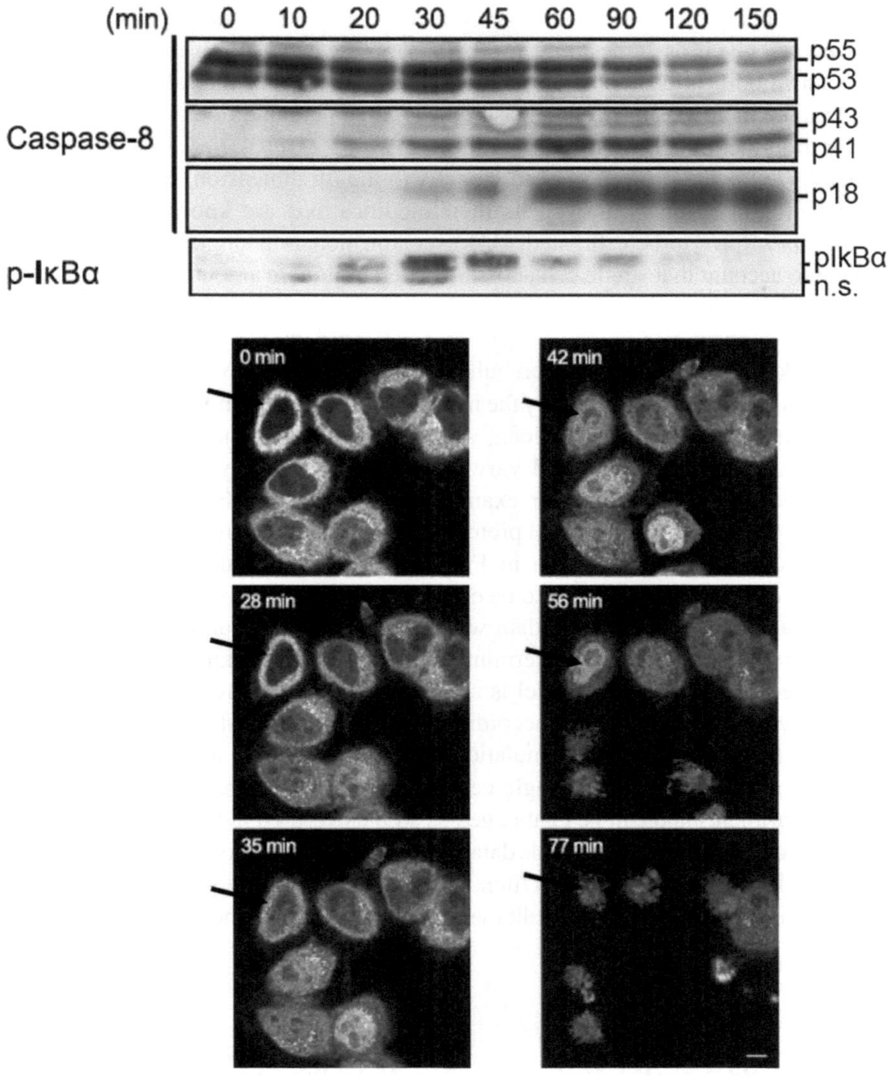

Fig. 7.7 Simultaneous activation of both NFκB and caspase-8. *Top*: Cleavage of caspase-8 and phosphorylation of IκBa as determined by WB upon stimulation of HeLa-CD95 cells with agonistic anti-APO-1 antibody. *Bottom*: Single-cell data of p65-mCherry transfected HeLa-CD95 cells upon stimulation with anti-APO-1. Nuclear localization of p65-mCherry precedes apoptosis. Modified from Neumann et al. (2010)

measurements of single cells have to be used. The protein amounts of each of these cells have to be guessed. This can be very time consuming.

In the following we will compare fitting of models based on bulk measurements to those based on single-cell data: We denote $\vec{c}(0)$ as the initial protein concentrations, t as the time, $\vec{s}$ as the experimental conditions (for example, stimulation), and $\vec{k}$ as the rate constants. A model $f : (\vec{c}(0), t, \vec{s}, \vec{k}) = \vec{c}$ yields protein amounts, modifications, or other outputs at a given time point upon stimulation $\vec{s}$. Measurement values are given as tuples $m_{i,j} = \langle \vec{s}_i, t_j, \vec{c}_{i,j} \rangle$. $\vec{C}_{i,j}$ denotes the measured protein amounts/modifications upon stimulation $\vec{s}_i$ and time t_j. For the sake of simplicity, we assume there was only one measurement value per condition and time point. In order to fit the model, we try to minimize: $\sum_{\forall i,j:\exists m_{i,j}} \left(f(\vec{c}(0), t_j, \vec{s}_i, \vec{k}) - p(3, m_{i,j})\right)^2$, where $p(i,x)$ is the projection of the tupel x to the ith dimension. In case the mean initial protein amounts as well as their modifications are known, only the rate constants $\vec{k}$ have to be estimated. However, for modeling single cells, we have to take into account that single cells possess different protein amounts. We will denote the initial protein concentrations of a cell n as $\vec{c}_n$. In order to fit the model, we would have to minimize: $\sum_{\forall n,i,j:\exists m_{i,j,n}} \left(f(\vec{c}_n(0), t_j, \vec{s}_i, \vec{k}) - p(4, m_{i,j,n})\right)^2$ with $m_{i,j,n} = \langle \vec{s}_i, t_j, n, \vec{c}_{i,j,n} \rangle$. Like for the model based on bulk measurements, we have to estimate the rate constants $\vec{k}$. However, additionally the initial protein concentrations or the initial state of the cells have to be estimated locally, too. Term "locally" means that these values are estimated for each cell and vary between cells. This type of task could be performed by software like, for example, "potterswheel." For large amounts of cells, the distribution of the initial protein amounts should approximate a log-normal distribution like the one shown in Fig. 7.1. In contrast, rate constants must be identical in all cells. They have to be estimated globally. In comparison to fitting to bulk data, fitting to single-cell data will be more time consuming and the model parameters might be less determined, as additional unknown variables are introduced. If the resulting model is simulated several hundred or thousand times with varying initial conditions according to a log-normal distribution as shown in Fig. 7.1, then the sum of the simulations outputs will mimic a bulk measurement. Therefore, a model based on single-cell data is capable of predicting the output of bulk experiments and can be challenged as well as verified in this way. However, a model based exclusively on bulk data will in many cases not be able to describe single-cell behavior accurately. Often, however, the topology which was identified when constructing a model on bulk measurements might be the same.

7.5 Caspase-3 Activation in Type II Cells: A Case Study

An example of how single-cell measurements can be applied in systems biology is the modeling of activation of caspase-3 in type II cells. Groups of Jochen Prehn and Peter Sorger made a significant contribution to unravel this question. Upon DR stimulation the death-inducing signaling complex (DISC) is formed. Procaspase-8 is activated at the DISC. Following procaspase-8 activation at the DISC, apoptotic signaling can go via type I or type II, which is described in detail in Chap. 5. Type I cells are characterized by high levels of DISC formation and high amounts of active caspase-8. Activated caspase-8 directly leads to activation of downstream effector caspases-3 and -7. In type II cells, there are lower levels of DISC formation and, thus, lower levels of active caspase-8. In this case, comprising the majority of cell lines, signaling requires an amplification loop. This amplification loop involves the cleavage by caspase-8 of the Bcl-2 family protein Bid to generate truncated (t) Bid and subsequent tBid-mitochondrial outer membrane permeabilization (MOMP). Only after mitochondrial outer membrane permeabilization (MOMP), caspase-3 is fully activated and apoptosis is initiated. This was shown using FRET reporter capable of measuring caspase-3 activity by Rehm et al. (2002). Upon cleavage of the reporter by caspase-3, the FRET signal is diminished as the interaction between the fluorophores of the probe is lost. Figure 7.8a shows caspase-3 activity as measured in a bulk experiment. Figure 7.8b displays the kinetic of FRET probe cleavage in single cells treated with TNF-α. In the bulk experiment it seems that there is a slow and steady increase over time. However, single-cell data reveal that the actual behavior is different. There is a lag time after stimulation with TNFα. In this time interval, no change in probe fluorescence can be detected. This lag time was also termed pre-MOMP delay. The duration of the lag time greatly varies between single cells. The lag phase is followed by a rapid cleavage of the probe. Within 10 min, complete processing of the probe takes place. This is independent of the stimulus strength used to induce MOMP (Fig. 7.8). The activation of caspase-3 follows an all-or-none behavior. Other groups made the same observation (Goldstein et al. 2000; Tyas et al. 2000). Either no substrate cleavage occurs or all substrates are rapidly cleaved. Interestingly MOMP preceded activation of caspase-3 by 10 min.

Rehm and colleagues found that probe processing could be well described by the following equation: $c(t) = f - \frac{f}{1+e^{(t-Td)/4Ts}}$.

In this equation $c(t)$ is the amount of cleaved substrate at time t, f is the fraction cleaved at the end of the reactions, and Td is the delay between stimulation and the half-maximal substrate cleavage. Ts is the switching time between initial and complete effector substrate cleavage. In 2006 the same group combined the caspase-3 activity probe measurements with tetramethyl rhodamine ethyl ester (TMRM), a marker of the mitochondrial potential. Based on these data, a model describing caspase-3 activation upon MOMP was built (Fig. 7.9). Applying sensitivity analysis, the authors identified procaspase-9, -3, and the XIAP concentrations as the main factors which determine rapid caspase-3 activation following MOMP. There are two reasons for the all-or-none behavior of caspase-3. First, cleaved and thereby activated caspase-3 is counteracted by XIAP. Hence, when the amount of the active caspase-3 is above the one that could be blocked by XIAP, the caspase-3 activation could not be blocked anymore. Second, as soon as cyt c is released, the apoptosome is being formed, and caspase-9 is activated. A positive feedback between caspase-3 and caspase-9 is turned on, which leads to rapid caspase-3 activation. The authors predicted that MOMP-induced apoptosis initiation will be slowed down upon XIAP overexpression and could confirm this prediction experimentally. Moreover, overexpression of XIAP could lead to sublethal caspase activation and incomplete substrate cleavage. This has been proposed to promote tumor formation.

Two years later the group of Peter Sorger came up with a larger model describing both caspase-8 and subsequent caspase-3 activation upon TRAIL stimulation (Fig. 7.10) (Albeck et al. 2008b). This was the first model based on single-cell data

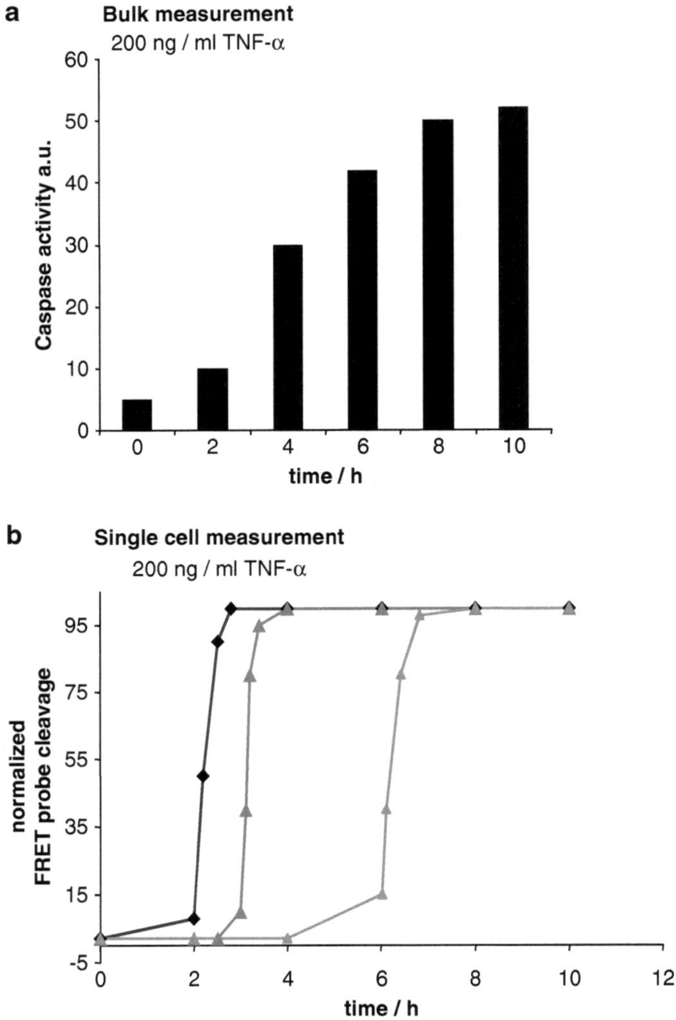

Fig. 7.8 Kinetics of effector caspase activation upon TNF-α stimulation. (**a**) Caspase-3 activity as determined by DEVD cleavage upon stimulation of HeLA cells with TNF-α plus cycloheximide. Bulk measurements suggest a gradual increase in activity. (**b**) Caspase-3 activity as determined in single cells with a FRET probe. Processing of the FRET probe occurs rapidly after a lag phase of variable length. Modified from Rehm et al. (2002)

combining both the extrinsic and the intrinsic pathways. The authors used probes which could specifically be cleaved by either caspase-8 (termed initiator caspase reporter protein, IC-RP) or caspase-3/7 (termed effector caspase reporter protein, EC-RP). Using the caspase-8-specific probe, it was demonstrated that caspase-8 activity gradually builds up at the time before MOMP termed pre-MOMP. The onset of caspase-3 activity, however, occurred after cyt c release (Fig. 7.11a)

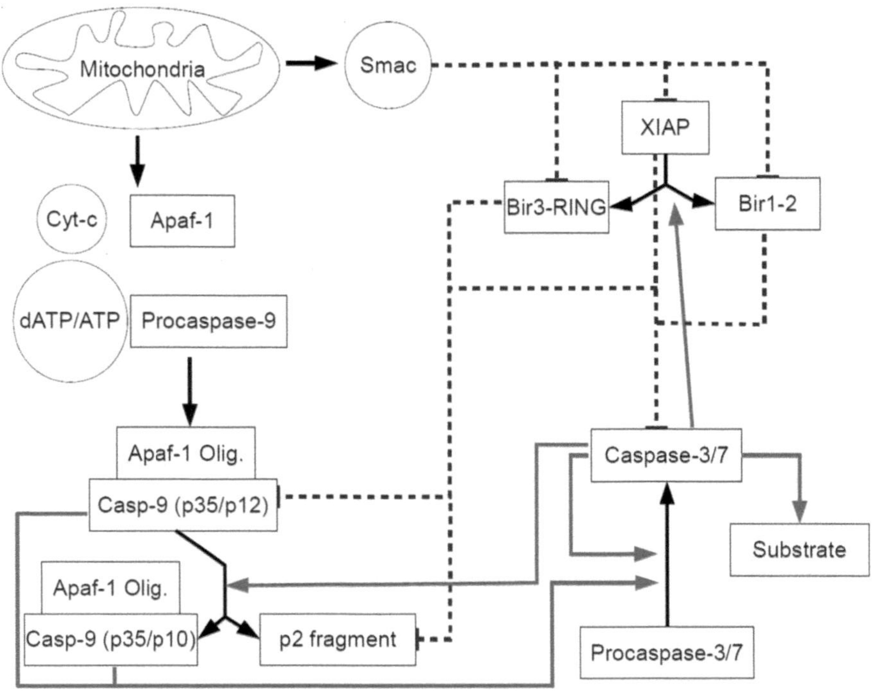

Fig. 7.9 A model of caspase-3/-9 activation upon MOMP. A scheme of a model of caspase-3 and -9 activation following MOMP. Enzymatic cleavage is indicated by *green arrows*. Inhibitory reactions are shown as *red lines*. Modified from Rehm et al. (2006)

(Albeck et al. 2008a). The variable delay in the onset of caspase-3 activity among single cells could be mainly explained by differences in initial caspase-8 activity among single cells. In addition, Smac was identified as an important regulator of rapid caspase-3 activation. As soon as MOMP occurs, Smac is released from the mitochondria leading to XIAP degradation, thereby allowing caspase-3 activation. In a series of experiments, the authors could confirm the proposed mechanism of a rapid switch from inactive to active caspase-3, which the authors termed "snap-action of caspase-3" activation:

1. The authors proposed that procaspase-3 was cleaved before MOMP, however, rapidly inactivated by XIAP. Therefore, XIAP depletion should abolish this inhibition and allow effector caspase activity before MOMP. Indeed, simultaneous depletion of both Bid (in order to prevent MOMP) and XIAP resulted in a gradual increase in EC-RP cleavage upon stimulation (Fig. 7.11b).
2. Upon MOMP Smac is released from the mitochondria leading to XIAP degradation. This would abolish the caspase-3 inhibition by XIAP and contribute to snap-action of caspase-3 activation. The Albeck model predicted that upon Smac down-regulation, caspase-3 activity would increase more slowly following

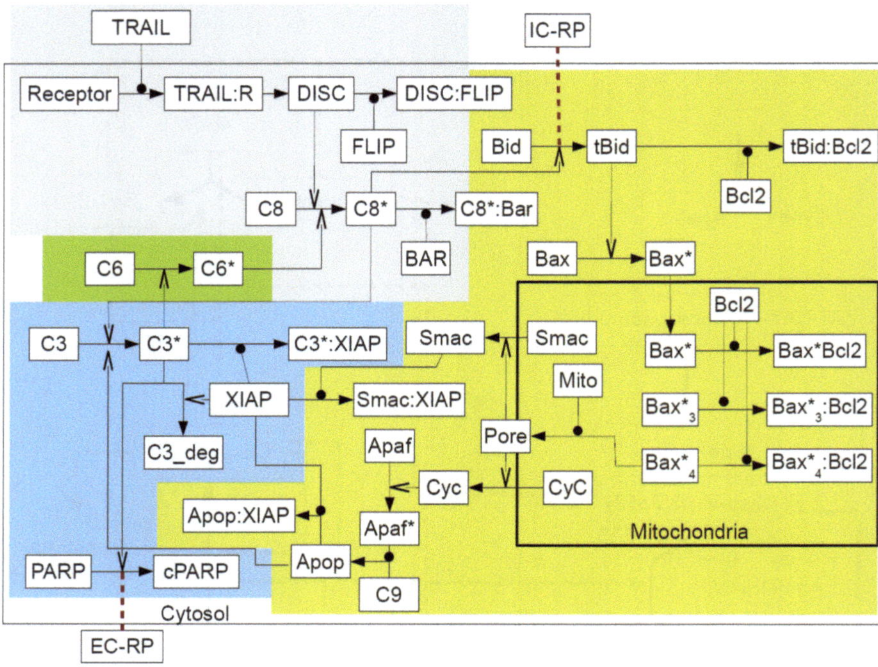

Fig. 7.10 A model of the TRAIL signaling pathway. A scheme of the TRAIL signaling pathway. *Symbols* in the scheme follow the convention of Kitano et al. *Background colors* indicate different network modules. Modified from Albeck et al. (2008b)

MOMP. This prediction could be confirmed experimentally: Upon Smac knock-down, a gradual increase in effector caspase activity occurred (Fig. 7.11c).

The Albeck model was able to reproduce both single-cell and WB data accurately. It consists of 70 rate constants and 98 species. Interestingly, the model parameters were taken from the literature as well as fitted manually. The model nicely proposed a mechanism for the all-or-none behavior observed with respect to caspase-3 activation. All-or-none behavior of effector caspase activation is an important property, as it establishes MOMP as a point of no return. MOMP leads to full activation of caspases-3 and -9 and cell death.

7.6 Investigation of Nongenetic Cell-to-Cell Variability in the TRAIL Pathway

Spencer et al. (2009) investigated nongenetic cell-to-cell variability. The authors found that the phase within the cell cycle or cell size barely influences the mean time to MOMP upon TRAIL stimulation (Spencer et al. 2009). However they discovered that sister cells shortly after cellular division behaved very similarly. Similarity

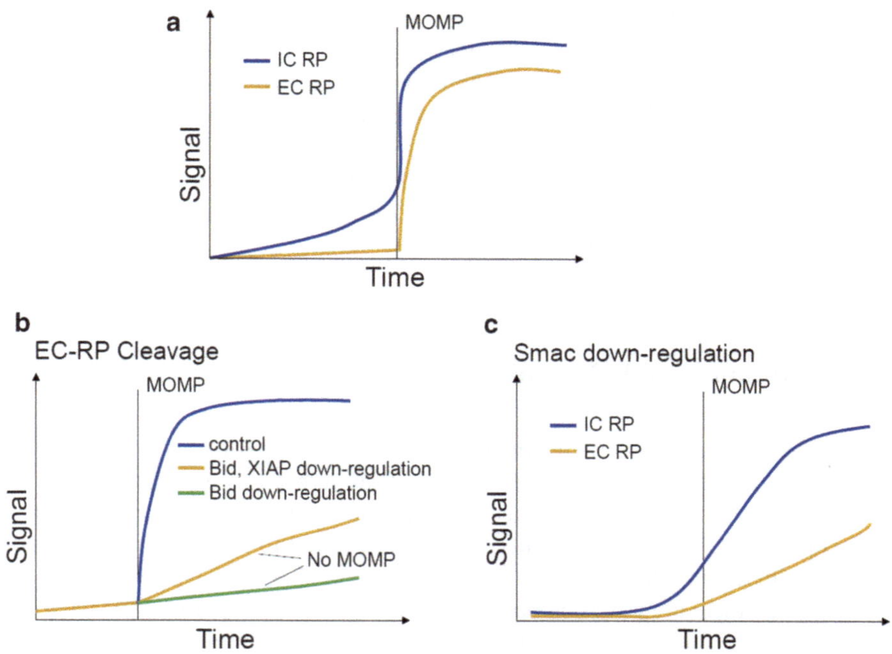

Fig. 7.11 Snap-action of caspase-3. (**a**) Cleavage of IC-RP and EC-RP in HeLa cell treated with 50 ng/ml TRAIL plus 2.5 μg/ml cycloheximide. Caspase-8 activity (IC-RP) gradually increases until MOMP. Caspase-3 activity (EC-RP) is absent before MOMP. Cells are aligned by the average time of MOMP. (**b**) EC-RP cleavage upon down-regulation of Bid or Bid plus XIAP. Bid down-regulation will prevent MOMP. Upon co-down-regulation of XIAP, a gradual increase in probe cleavage is observed. (**c**) Cleavage of EC-RP and IC-RP upon Smac down-regulation. Upon down-regulation of Smac, snap-action of caspase-3 is abolished. Instead a gradual and slow increase in caspase-3 activity is observed. Modified from Albeck et al. (2008b)

between sister pairs decreased with increasing time after cellular division. Directly after division two daughter cells have almost identical protein amounts and modifications. It was suggested that stochastic events during transcription could lead to different protein levels among two daughter cells. These differences would accumulate. Several hours after cellular division, two sister cells will not be more similar than any two random cells. The authors therefore proposed that random differences in protein amounts are one major source of cell-to-cell variability. The model as published by Albeck et al. (2008b) could indeed reproduce this observation. Differences in protein expression of a factor of 2.5 were sufficient to cause differences in phenotype. Another important prediction was that for the TRAIL pathway, there is not one single most important protein deciding the outcome of stimulation. If, for example, the amount of procaspase-8 is relatively high in a single cell, this does not mean that this cell is especially sensitive towards TRAIL-induced apoptosis. As there is a number of anti-apoptotic players involved, the pro-apoptotic effect of procaspase-8 might be compensated by these proteins. This prediction could be elegantly confirmed by Spencer and coauthors. A cell line was used expressing a c-FLIP YFP fusion protein at the endogenous locus, resulting in physiological expression levels of the construct (Spencer et al. 2009). Surprisingly, no correlation was found between the expression of c-FLIP YFP in single cells and the time when MOMP occurred after TRAIL stimulation. On the basis of these observations, it was suggested that variations in the expression level of one protein are not sufficient to determine sensitivity. Whether other sources of cell-to-cell variability like, for example, protein phosphorylation might have similar impacts is an issue to be addressed in the future.

7.7 Summary and Outlook

In the previous sections, we learned how single-cell experiments could provide fascinating insights into cellular processes. They allow to investigate dynamics which cannot be detected in bulk measurements. In combination with mathematical modeling, single-cell measurements are a powerful tool to gain better understanding in the field of biology.

However, there are issues on both the modeling and the experimental side to be improved in the future. So far it has not been possible to monitor a large number of events simultaneously in a single cell. Especially FRET probes have the disadvantage that they occupy a large area of the wavelength spectrum, naturally restricting a number of proteins to be monitored with FRET probes. Measuring a larger number of cellular events in parallel would allow investigating the interplay of signaling processes. On the modeling side mostly ODEs have been applied so far with great success (Bentele et al. 2004; Albeck et al. 2008a, b; Neumann et al. 2010; Fricker et al 2010). However, as Spencer et al. (2009) showed that random fluctuations in transcription have a large impact on cellular behavior, stochastic modeling will be more and more important in the future. Especially in models in which transcription has to be incorporated, stochastic models might describe the cellular behavior more accurately. Taken together, further development of both experimental methods and modeling approaches could provide even more exciting insights into regulation of cellular decisions.

Acknowledgments We acknowledge the Helmholtz Alliance on Systems Biology (NW1SBCancer) and Helmholtz-Russia Joint Research Groups-2008-2 for supporting our work.

References

Albeck JG, Burke JM, Aldridge BB et al (2008a) Quantitative analysis of pathways controlling extrinsic apoptosis in single cells. Mol Cell 30:11–25. doi:10.1016/j.molcel.2008.02.012

Albeck JG, Burke JM, Spencer SL et al (2008b) Modeling a snap-action, variable-delay switch controlling extrinsic cell death. PLoS Biol 6:2831–2852. doi:10.1371/journal.pbio.0060299

Bentele M, Lavrik I, Ulrich M et al (2004) Mathematical modeling reveals threshold mechanism in CD95-induced apoptosis. J Cell Biol 166:839–851. doi:10.1083/jcb.200404158

Fricker N, Beaudouin J, Richter P et al (2010) Model-based dissection of CD95 signaling dynamics reveals both a pro- and antiapoptotic role of c-FLIPL. J Cell Biol 190:377–389. doi:10.1083/jcb.201002060

Goldstein JC, Waterhouse NJ, Juin P et al (2000) The coordinate release of cytochrome c during apoptosis is rapid, complete and kinetically invariant. Nat Cell Biol 2:156–162. doi:10.1038/35004029

Loewer A, Batchelor E, Gaglia G, Lahav G (2010) Basal dynamics of p53 reveal transcriptionally attenuated pulses in cycling cells. Cell 142:89–100. doi:10.1016/j.cell.2010.05.031

Neumann L, Pforr C, Beaudouin J et al (2010) Dynamics within the CD95 death-inducing signaling complex decide life and death of cells. Mol Syst Biol 6:352. doi:10.1038/msb.2010.6

Raj A, van Oudenaarden A (2008) Nature, nurture, or chance: stochastic gene expression and its consequences. Cell 135:216–26. doi:10.1016/j.cell.2008.09.050

Rehm M, Dussmann H, Janicke RU et al (2002) Single-cell fluorescence resonance energy transfer analysis demonstrates that caspase activation during apoptosis is a rapid process. Role of caspase-3. J Biol Chem 277:24506–24514. doi:10.1074/jbc.M110789200

Rehm M, Huber HJ, Dussmann H, Prehn JHM (2006) Systems analysis of effector caspase activation and its control by X-linked inhibitor of apoptosis protein. EMBO J 25:4338–4349. doi:10.1038/sj.emboj.7601295

Schilling M, Maiwald T, Bohl S et al (2005) Computational processing and error reduction strategies for standardized quantitative data in biological networks. FEBS J 272:6400. doi:10.1111/j.1742-4658.2005.05037

Spencer SL, Sorger PK (2011) Measuring and modeling apoptosis in single cells. Cell 144:926–39. doi:10.1016/j.cell.2011.03.002

Spencer SL, Gaudet S, Albeck JG et al (2009) Non-genetic origins of cell-to-cell variability in TRAIL-induced apoptosis. Nature 459:428–32. doi:10.1038/nature08012

Tyas L, Brophy VA, Pope A et al (2000) Rapid caspase-3 activation during apoptosis revealed using fluorescence-resonance energy transfer. EMBO Rep 1:266–270. doi:10.1093/embo-reports/kvd050

Chapter 8
Cytokine–Cytokine Cross Talk and Cell-Death Decisions

Christopher D. Deppmann and Kevin A. Janes

Abstract Cells often face life or death decisions in response to conflicting extracellular cytokine cues. A full understanding of how this information is encoded has implications for normal development and function of organ systems and also for pathologies where cues are not processed properly. In this chapter, we discuss how life–death decisions are influenced by signaling from pro-survival receptor tyrosine kinases (RTKs) and pro-death tumor necrosis factor (TNF)-family receptors. Intracellular cross talk between these antagonistic receptors is incredibly complex, and our understanding could be improved by systems biology thinking and approaches. We describe key systems-level features of RTK and TNF-family receptor signaling and how points of cross talk may mediate the decision to live or die.

8.1 Introduction

Cells live in a world of mixed messages. Many cellular instructions come from the microenvironment, which provides cues that are diverse and often contradictory: proliferate *and* quiesce, differentiate into one lineage *and* another, etc. How do cells process such information and respond appropriately? This question is particularly important when cell numbers and organization must be precisely controlled, as in tissues with rapid turnover and during development.

For each cell, no choice is more critical or irreversible than the decision to die. Consequently, cell-death decisions are steered by many factors—adhesion to

C.D. Deppmann (✉)
Department of Biology, University of Virginia, Charlottesville, VA 22908, USA
e-mail: deppmann@virginia.edu

K.A. Janes (✉)
Department of Biomedical Engineering, University of Virginia, Charlottesville, VA 22908, USA
e-mail: kjanes@virginia.edu

I.N. Lavrik (ed.), *Systems Biology of Apoptosis*, DOI 10.1007/978-1-4614-4009-3_8,

neighboring cells and the extracellular matrix (ECM; see Table 8.1 for a complete list of abbreviations), cell cycle status, age, and environmental insults, among others. One prominent source of information comes from *cytokines*, a term we use here to refer collectively to all diffusible protein ligands (Janeway et al. 2001). Some cytokines, such as those of the tumor necrosis factor (TNF) superfamily, induce apoptosis in many cell types. Others, such as cytokines from the epidermal, insulin-like, and nerve growth factor (EGF, IGF, and NGF) families, act as survival cues that block or attenuate the action of TNF-family cytokines through the action of their receptor tyrosine kinases (RTKs). Pro-death and pro-survival cytokines intersect at many points inside and outside the cell, and their cross talk is the focus of this chapter.

Cross-communication between conflicting cytokines underlies many normal and pathological functions. During neural development, competition between death signals from the p75-NGF receptor and survival signals from Trk RTKs defines the number of neurons innervating a target tissue (Deppmann et al. 2008; Majdan et al. 2001). In a particular neuron, one pathway appears to become dominant over the other, ultimately leading to the life–death decision. This seems to depend on the strength of RTK signaling. For example, in cultured sympathetic neurons, if NGF concentrations are high enough for "optimal" activation of TrkA, no amount of a TNF-family ligand can induce death even when added at saturating concentrations (Deppmann et al. 2008; Kohn et al. 1999). On the other hand, when NGF is present below this threshold (~5 ng/ml), TNF-family receptor signaling can become dominant by suppressing survival signaling from RTKs and inducing apoptosis (Bamji et al. 1998). In many non-neuronal cell types, the pro-apoptotic functions of TNF are antagonized by EGF (Akca et al. 2003; Garcia-Lloret et al. 1996; Gibson et al. 1999). Likewise, insulin and IGFs have been shown to inhibit many TNF-induced responses (Bedard et al. 1998; Goetze et al. 2001; Qian et al. 2001; Weiner et al. 1991; Wu et al. 1996). For these types of antagonistic stimuli, communication between signaling pathways through cross talk is critical in determining outcome.

Cytokine–cytokine cross talk is important for normal cellular functions during embryonic development and homeostasis in the adult. However, these pathways are often misregulated and give rise to a wide range of pathologies. In colonic epithelia, for example, TNF is a key mediator of inflammatory bowel disease (Rutgeerts et al. 2004), whereas EGF and IGF stimulate growth and repair of the mucosa (Chailler and Menard 1999; Singh and Rubin 1993). A similar TNF-mediated inflammatory response is thought to critically influence the rate of neurodegeneration during Alzheimer's disease (Perry et al. 2001).

The imbalance between pro-death and pro-survival cytokine signaling is perhaps best documented in cancer. Various cancers amplify pro-growth signals and suppress pro-apoptotic signals (Hanahan and Weinberg 2000, 2011). How these pathways impinge on one another in the context of tumorigenesis is the subject of ongoing research. Much progress has been made toward understanding how cross talk pathways are co-opted during cancer. For extrinsic death signaling, loss-of-function mutations have been reported in tumors for several TNF-family

Table 8.1 List of abbreviations used

ADAM	A disintegrin and metalloproteinase
AP-1	Activator protein-1
Bcl2	B-cell lymphoma protein 2
c-IAP	Cellular inhibitor of apoptosis protein
DcR3	Decoy receptor 3
DISC	Death-inducing signaling complex
DR4	Death receptor 4
DR5	Death receptor 5
ECM	Extracellular matrix
EGF	Epidermal growth factor
ERK	Extracellular regulated kinase
FGF	Fibroblast growth factor
FLIP	FLICE-like inhibitor protein
FOXO	Forkhead box O
IGF	Insulin-like growth factor
IL-1	Interleukin-1
IL-1ra	IL-1 receptor antagonist
IKK	IκB kinase
IRF	Interferon response factor
IRS	Insulin receptor substrate
JNK	c-jun N-terminal kinase
MAPK	Mitogen-activated protein kinase
MKP1	MAP kinase phosphatase 1
NFκB	Nuclear factor κB
NGF	Nerve growth factor
PI3K	Phosphoinositide 3-kinase
PIP2	Phosphatidylinositol 3,4-bisphosphate
PIP3	Phosphatidylinositol 3,4,5-trisphosphate
PTB	Phosphotyrosine binding
PTEN	Phosphatase and tensin homolog
RIP1	Receptor interacting protein 1
RTK	Receptor tyrosine kinase
SH2	Src homology 2
TCF/LEF	T-cell factor/lymphoid enhancer factor
TGFα	Transforming growth factor-α
TLR	Toll-like receptor
TNF	Tumor necrosis factor
TRADD	TNF receptor-associated death-domain protein
TRAF2	TNF receptor-associated factor 2
TRAIL	TNF-related apoptosis-inducing ligand
TRID	TRAIL receptor without an intracellular domain
TRK	Tropomoysin receptor kinase
XIAP	X-linked inhibitor of apoptosis protein

receptors, including Fas and the death receptors 4 and 5 (DR4, DR5) (Ozoren and El-Deiry 2003). Cancer cells further inactivate death signaling by overexpression of decoy receptors, such as DcR3 and TRID (Pitti et al. 1998; Sheikh et al. 1999), or overexpression of downstream inhibitors of apoptosis, such as Bcl2, FLIP, XIAP, c-IAP, and survivin (Altieri 2003; LaCasse et al. 1998). Cancers have also been shown to mimic RTK-mediated survival signaling through several mechanisms. For example, the EGF receptor family member, ErbB2, is upregulated in roughly 30% of all breast cancers and is a clinical target for pharmacological interventions such as Herceptin (Bange et al. 2001). In addition, many downstream RTK effector pathways are upregulated or acquire gain-of-function mutations in cancer (Blume-Jensen and Hunter 2001). Interestingly, several commonly mutated oncogenes and tumor suppressors such as Akt or NFκB (Franke et al. 1997; Perkins 2004) lie at points of cross talk for survival and death signaling (discussed later in the chapter). A quantitative understanding of how these pathways interact promises to yield insight into how we might exploit these pathways to reinstate a surveillance system that promotes tumor regression.

Both normal and pathological instances of cytokine–cytokine cross talk are far more complex than one signal from a given RTK interacting with one signal of a particular TNF-family receptor. Cells usually express several members of both receptor classes. Aside from receptor expression, we must take into account where and in what manner the cell encounters death–survival ligands, how those ligands are processed, and how autocrine, paracrine, or endocrine interactions are interpreted. Moreover, it will be critical to understand the molecular mechanisms by which a survival pathway becomes dominant over a death pathway and vice versa.

Systems-level analysis is well suited to delineate the complexities of cytokine–cytokine cross talk. First, the phenomenon of cross talk is clearly an emergent property, which cannot be approached by studying single cytokines in isolation (Bhalla and Iyengar 1999). This requires multiparametric approaches that are designed to handle multiple inputs (Chatterjee et al. 2010; Garmaroudi et al. 2010). Second, various lines of evidence suggest that cytokine-induced signaling pathways transmit quantitative information en route to cell-fate decisions (Cohen-Saidon et al. 2009; Janes et al. 2005, 2008; Miller-Jensen et al. 2007). There are elements of qualitative logic (Aldridge et al. 2009; Janes et al. 2006; Zhang et al. 2008), but the most predictive models are those based on quantitative signal activation and propagation (Gordus et al. 2009; Janes et al. 2008; Kumar et al. 2007). Thus, biochemical measurements that are truly quantitative must be obtained to constrain these types of systems models adequately (Albeck et al. 2006; Janes and Yaffe 2006).

In this chapter, we focus on the systems-level cross talk between apoptotic TNF-family cytokines and pro-survival cytokines that directly activate RTKs. We begin with a brief introduction of TNF-family receptor signaling and RTK signaling, focusing on general recurring themes rather than the details of specific receptors. The bulk of the chapter is then organized around two major categories of cytokine–cytokine cross talk: (1) intracellular cross talk through dynamic transcriptional, and posttranslational modifications and (2) extracellular cross talk through constitutive or regulated autocrine circuits that reconfigure the microenvironment. We restrict our material to cross talk *between* cytokines, omitting any discussion of the cross talk that is known to occur between pathways activated by TNF-family or pro-survival cytokines on their own [for reviews on these topics, see

Avraham and Yarden (2011); Ware (2005)]. Last, we conclude by highlighting unanswered questions in cytokine–cytokine cross talk that could be tackled in the future with systems biology approaches.

8.2 Canonical Signaling Pathways Activated by TNF-Family Receptors and RTKs

Signal transduction is the process by which an extracellular cue is converted to an intracellular response (Downward 2001). A common requirement for both TNF-family receptors and RTKs to transduce signals is receptor oligomerization. Upon receptor clustering and activation, the active receptors recruit effector molecules to elicit immediate downstream posttranslational responses, which ultimately lead to changes in gene expression and a cell's molecular state. Below, we outline these basic steps of signal transduction in conceptual and molecular terms.

8.2.1 Receptor Oligomerization

Cytokine receptor activation is conventionally thought to occur when a multimeric ligand causes the dimerization or oligomerization of monomeric receptors to trigger downstream signal transduction. For RTKs, bivalent ligands were believed to bring two RTK monomers into close proximity of each other, allowing for trans-phosphorylation of intracellular domains and recruitment of downstream effectors (Ullrich and Schlessinger 1990). In contrast to RTKs, TNF-family receptors do not have enzymatic activity and so signal transduction is likely initiated by a conformational change (Locksley et al. 2001). This view of receptor activation is likely to be true in many cases, but structural studies have revealed several novel mechanisms by which ligands induce signaling independent from receptor oligomerization (Lemmon and Schlessinger 2010). For example, Trk and Kit RTKs require ligand to provide a dimerization interface, consistent with early models of RTK activation (Wehrman et al. 2007; Yuzawa et al. 2007). But for the EGF receptor and the fibroblast growth factor (FGF) receptor, ligand binding is completely dispensable for receptor dimerization (Huse and Kuriyan 2002; Schlessinger 2000). Instead, ligand–receptor interactions likely induce the conformational change of a preformed dimer, which in turn causes apposition of the intracellular kinase domains and trans-phosphorylation. Likewise, for some TNF-family receptors, such as p75-NGF receptor, dimers are thought to pre-exist. By virtue of a reciprocal "hinge" at the dimer interface, the two intracellular domains move apart from one another when ligand binding brings the extracellular domains closer together (Vilar et al. 2009).

We have only a partial understanding of the dynamics of ligand-induced receptor oligomerization and the accompanying conformational changes, precluding detailed models of how distinct cytokine pathways interact with each other at the receptor level.

8.2.2 *Effector Recruitment and Downstream Signaling*

Upon ligand binding to either TNF-family receptors or RTKs, combinations of effector proteins are recruited to the receptor's intracellular domain that ultimately determine the downstream pathways that will be activated. For RTKs, much of the binding is through protein interaction modules, such as SH2 and PTB domains, that selectively bind phosphotyrosine-containing sequence motifs (Pawson and Scott 1997). Proteins containing these interaction domains can serve as adaptors that recruit other proteins with enzymatic activity, such as guanine exchange factors and kinases. This recruitment is a key step in defining the intracellular signals that emanate from RTKs.

Nearly all RTKs canonically activate the Ras-MAPK, phospholipase C, and phosphoinositide 3-kinase (PI3K) pathways (McKay and Morrison 2007). Activation of these pathways leads to a wide range of responses including calcium release, induction of immediate early genes, and phosphoryation of several growth- and survival-promoting proteins in the cytoplasm (Lemmon and Schlessinger 2010; Sheng and Greenberg 1990). If all RTKs activate the same pathways, then why are there 58 different RTKs encoded in the human genome (Manning et al. 2002)? One explanation is that RTKs vary widely in the strength and duration with which they activate the canonical pathways. Emerging mechanisms for these differences as well as the implications for death–survival decisions will be discussed later in this chapter.

In contrast to the shared intracellular tyrosine kinase domain common to all RTKs, the cytoplasmic domains of TNF-family receptors are not as uniform. This may not be surprising considering that the 27 TNF-family receptors in humans were grouped on the basis of their cysteine-rich extracellular domains rather than a common feature of their intracellular domain as for RTKs (Locksley et al. 2001). Thus, only 7–10 of the TNF-family receptors contain a canonical death domain that serves to recruit death-inducing effector complexes (Lavrik et al. 2005). For TNF-family receptors, the principle of effector recruitment is similar to RTKs. The death-inducing signaling complex (DISC) links the TNF-family receptor to initiator caspases 8 or 10, which become activated by cleavage to promote activation of effector caspase pathways, such as caspase 3 and 6, which rapidly dismantle the cell. Alternatively, recruitment of non-apoptotic complexes may occur, such TRADD, RIP1, TRAF2, and cIAP1. These binding events do not activate caspases and instead promote activation of transcriptional events through IKK–NFκB or JNK–AP-1 (Ashkenazi and Dixit 1998). Whether and how pro-survival RTK activation influences effector recruitment by TNF-family receptors remains an open question.

8.3 Intracellular Cross talk

Within minutes of receptor binding, TNF-family cytokines and RTK ligands cause posttranslational modifications on signaling proteins inside the cell (see above). The canonical signal transduction pathways activated by TNF-family receptors and RTKs may appear separable. However, in many contexts, TNF-activated pathways can modulate signal transmission through RTK-activated pathways and vice versa. Posttranslational cross talk is often specific to certain cell types, stimulus conditions, and organisms, suggesting that these arose relatively recently during evolution.

Phosphoinositide 3-kinase (PI3K) is a classic downstream RTK effector, which is recruited to the receptor and activated via SH2-dependent binding of the PI3K regulatory subunit. Active PI3K catalyzes the conversion of phosphatidylinositol 3,4-bisphosphate (PIP2) to phosphatidylinositol 3,4,5-trisphosphate (PIP3) in the plasma membrane, ultimately leading to activation of the serine-threonine kinase Akt (Datta et al. 1999). Akt pathway activity is negatively regulated by the tumor suppressor phosphatase and tensin homolog (PTEN), a lipid phosphatase that catalyzes the conversion of PIP3 back to PIP2.

Akt is also a target for cross talk from pathways activated by TNF-family receptors. One way in which TNF-family receptor signaling silences this pathway is through upregulation of PTEN. In neurons, p75-NGF receptor mediates PTEN activation, inhibiting PI3K and Akt signaling (Cantley and Neel 1999; Song et al. 2010). This may be clinically relevant for neuron loss in the context of stroke, where inhibition of PTEN prevents p75-NGF receptor-mediated cell loss (Song et al. 2010).

Active Akt phosphorylates proteins that contain an RXRXX(S/T) consensus sequence (Obata et al. 2000), and Akt signaling is an important conduit for cross talk between TNF-family receptors and RTKs. Among stress and apoptotic pathways activated by TNF-family ligands, Akt phosphorylation events are usually inhibitory for cell death. For example, expression of the TNF-family ligand *TRAIL* is partly controlled by FOXO transcription factors, which are excluded from the nucleus when phosphorylated by Akt (Brunet et al. 1999; Modur et al. 2002). The pro-apoptotic Bcl2-family protein Bad contains an Akt consensus sequence. It has been shown that Akt phosphorylates BAD at this site, which inhibits permeabilization of the mitochondrial outer membrane and prevents apoptosis (Datta et al. 1997; Scaffidi et al. 1998). Interestingly, *BAD S3A* knock-in mice (where all Akt phosphorylation sites have been mutated to alanines) display only a mild apoptotic phenotype (Datta et al. 2002). This suggests that Akt–Bad cross talk may be more critical for the metabolic functions of Akt than for controlling apoptosis (Danial et al. 2003).

Akt has been reported to phosphorylate other proteins activated by TNF-family cytokines, but their functional importance is more controversial. For instance, Akt can directly activate NFκB through phosphorylation of IKKα (Ozes et al. 1999). However, IKKα is not essential for NFκB activation, and relative importance of

Akt–NFκB cross talk appears to be highly cell-type specific (Delhase et al. 2000; Gustin et al. 2004; Li et al. 1999). This pathway may nevertheless be most important in cancers with PTEN loss, which have been found to have constitutively high NFκB activity (Fernandez-Marcos et al. 2009). Another inhibitory phosphorylation catalyzed by Akt occurs on human caspase-9, an initiator caspase for apoptosis (Cardone et al. 1998). Here, the generality of this cross talk has been questioned, because the serine targeted by Akt in human caspase-9 is not conserved in other mammalian orthologs (Rodriguez et al. 2000). Other inhibitory phosphorylation sites on caspase-9, such as Thr125 catalyzed by ERK, show incomplete conservation across mammals (Allan et al. 2003). Identification of highly conserved sites for TNF receptor–RTK cross talk would imply that cells have frequently encountered dual-cytokine stimulation over the course of evolution (Janes 2010).

8.4 Modeling Posttranslational Cross Talk

The earliest efforts aimed at modeling cytokine–cytokine posttranslational cross talk were based entirely on quantitative measurements. In an observational study of signaling synergy between cytokines, Natarajan et al. found that strongly nonadditive cross talk between cytokines was infrequent but non-negligible (Natarajan et al. 2006). Later work by this group showed that these pairwise synergies could entirely explain signaling responses to higher-order cytokine combinations (Hsueh et al. 2009), suggesting that nonadditive signal processing may stop at pairs (Janes 2010).

Shortly before the synergy work of Natarajan and coworkers, a group at MIT published a triplet of papers reporting a systematic survey and analysis of apoptotic and survival signaling triggered by TNF, EGF, and insulin (Gaudet et al. 2005; Janes et al. 2005, 2006). Using a dataset of nearly 8,000 signaling events, the authors built a partial least squares that accurately predicted ~1,500 apoptotic outcomes at different time points and under different stimulus conditions (Janes et al. 2005). For widespread predictive accuracy, cytokine combinations were required, because the model could not predict the observed two-stimulus signaling trajectories using only the single-treatment data. Moreover, an in-depth analysis of the validated model showed that model predictions were more robust to measurement noise when trained on cytokine combinations, as opposed to a single cytokine (Gaudet et al. 2005). Together, these studies illustrated the value of modeling cytokine combinations explicitly to further our understanding of cell-death control at the systems level.

More recently, multi-cytokine models of signaling have incorporated basic molecular rules and pathway connectivity to build predictive intracellular networks that reflect and expand upon known mechanisms. A simple way for encoding molecular logic and wiring is through Boolean networks, which flip proteins ON or OFF depending on the ON/OFF status ("membership class") of the upstream proteins that feed into the logic "gate." Saez-Rodriguez and coworkers assembled from the literature a Boolean network model of posttranslational cross talk activated by seven cytokines in hepatocytes (Saez-Rodriguez et al. 2009). The authors complemented their model with multiplex phospho-protein measurements and found that these data were crucial for eliminating irrelevant interactions and improving predictive power. Using the TNF–EGF–insulin dataset described above, Aldridge et al. trained a fuzzy logic model of signaling cross talk (Aldridge et al. 2009). Rather than hard-coded rules as in Boolean networks, fuzzy logic models have more flexible membership classes that enable smoother transitions between states. From published studies, the authors manually assembled a fuzzy logic network interconnecting the ten intracellular proteins measured in the earlier TNF–EGF–insulin studies (Gaudet et al. 2005; Janes et al. 2005, 2006). Based on the model's ability to capture the time course of signaling observed upon TNF, EGF, and insulin stimulation, the authors refined this network with additional logic gates that substantially improved predictive accuracy. These "logical" additions led to new predictions about the relationships between intracellular proteins, which could be explored through mechanistic experiments in the future.

It is challenging to model posttranslational cross talk between cytokines mechanistically, even when the individual stimuli have been extensively studied at the systems level. For physicochemical modeling, there are important bookkeeping issues when combining two cytokine-signaling networks, as shared molecular species must be properly accounted for during model fusion (Aldridge et al. 2006). Borisov et al. reported the first of such model fusions for the EGF and insulin RTKs networks (Borisov et al. 2009). Their results highlighted the critical role that adaptor proteins—such as Gab1, Grb2, and IRS—play in mediating cross talk between shared pathways (ERK and Akt) that are activated both by EGF and insulin. Interestingly, using data-driven modeling approaches, Gordus and coworkers showed that simple multi-linear models based on binding affinities may be sufficient to predict cross talk at the adaptor level (Gordus et al. 2009; Janes and Yaffe 2006). This raises the future possibility of building hybrid models that use data-driven approaches to link cytokines to adaptors and then physicochemical approaches to connect adaptors to downstream signaling outputs.

8.5 Transcriptional Cross Talk

The extensive transcriptional changes caused by TNF family and RTK signaling provide ample opportunities for these pathways to cross talk on slower time scales. The main driver of TNF-induced gene expression is NFκB (Karin and Ben-Neriah 2000), and NFκB has been shown to cross-communicate with other transcriptional modulators. NFκB works together with interferon response factors (IRFs) to coordinate the expression of shared and stimulus-specific transcriptional targets (Cheng et al. 2011). Persistent NFκB activity is indirectly antagonized by activator protein-1 (AP-1) transcription factors during Toll-like receptor (TLR) signaling (Kim et al. 2005). NFκB-mediated transcription can also be directly repressed by β-catenin in breast and colon cancers, where β-catenin levels are abnormally high (Deng et al.

2002). Many AP-1 subunits are rapidly upregulated by RTK signaling (Amit et al. 2007), and TCF/LEF transcriptional activity through β-catenin has been shown to be responsive to growth factors (Graham and Asthagiri 2004). Thus, it would be interesting to examine how acute AP-1 and TCF/LEF activation converge with acute NFκB activation, as would be the case in cells costimulated with RTK-activating and TNF-family cytokines.

A secondary source of transcriptional cross talk occurs through the function of the target genes that are upregulated or repressed. For example, NFκB induces transcription of the phosphatase *MKP1*, which dephosphorylates MAP kinases such as ERK (Wang et al. 2008; Wu et al. 2006). Conversely, NFκB downregulates expression of the lipid phosphatase *PTEN*, leading to increased activation of Akt (Kim et al. 2004). Various cytokines are themselves subject to joint transcriptional regulation by TNF family and RTK signaling. For instance, the apoptotic TNF-family cytokine *FASL* is upregulated by the combined actions of NFκB and AP-1 (Kasibhatla et al. 1998). Upregulation of cytokine gene expression is one of several ways in which inducible autocrine circuits can become activated (see below).

8.6 Extracellular Cross Talk

With few exceptions, TNF-family ligands are first presented as bioactive membrane-bound oligomers, which in many instances can be converted into a soluble form by a transmembrane ADAM metallopeptidase, ADAM17 (also known as tumor necrosis factor-α-converting enzyme) (Idriss and Naismith 2000). This cleavage event converts the ligand from a principally juxtacrine mechanism between adjacent cells to a paracrine–autocrine mechanism of signaling. This has important functional consequences as it is thought that paracrine signaling via soluble TNF is critical for chronic inflammation (Holtmann and Neurath 2004; Ruuls et al. 2001), whereas juxtacrine signaling via membrane-bound TNF is more important for maintaining immunity to pathogens and resolving inflammation (Alexopoulou et al. 2006; Canault et al. 2004; Mueller et al. 1999). These different modes of signaling represent a common route to inter-cytokine cross talk through dual autocrine–paracrine signaling. Different classes of cytokines can become interlinked when one is secreted by a neighboring cell and the other is constitutively or inducibly released by the receiving cell. This type of extracellular cross talk is well suited to systems analyses, because autocrine circuits are highly iterative and time dependent. Thus, autocrine cytokines require both quantitative models and quantitative experiments to appreciate them fully (DeWitt et al. 2001; Lauffenburger et al. 1995; Monine et al. 2005).

Many TNF-family cytokines have been shown to participate in extracellular signaling circuits that engage other classes of cytokines. In response to TLR stimulation, TNFα is upregulated and released as an autocrine factor after a time delay for biosynthesis (Covert et al. 2005; Werner et al. 2005). Interestingly, modeling and experiments showed that this wiring (TLR → TNFα with a time delay) was critical for sustained NFκB activity. TNFα stimulation alone or TLR stimulation without autocrine TNFα led to damped oscillations in NFκB activity, which perturbed NFκB-dependent transcriptional responses compared to when the native extracellular circuit was intact.

In epithelia, TNF triggers its own cascade of autocrine cytokines that feed back to influence cell-fate responses. At early times after cytokine stimulation, TNF causes release of transforming growth factor-α (TGFα), which signals through RTKs to activate ERK (Chen et al. 2004; Janes et al. 2006). At later times, TNF transcriptionally upregulates *IL1A*, which contributes to sustained IKK–NFκB signaling downstream of TNF (Janes and Lauffenburger 2006; Janes et al. 2008). Remarkably, autocrine IL-1α signaling requires both the TNF stimulus and the TGFα autocrine factor for maximal signaling. Finally, at very late times, TNFα attenuates IL-1α signaling by inducing inhibitory autocrine signaling with IL-1ra, an IL-1 receptor antagonist. Together, these results (derived from systems-level signaling datasets and quantitative analyses) indicate a clear molecular logic among extracellular autocrine circuits (Janes et al. 2006). The basic elements of the TNFα–TGFα–IL-1α–IL-1ra cascade were shown in multiple cell types and later extended to control of proliferation and apoptosis in hepatocytes (Cosgrove et al. 2008). Intriguingly, the cytokines were the same, but the molecular logic interconnecting them was found to be different, suggesting a mechanism for conferring cell-type specificity to the same initial stimulus.

A largely unexamined area of research is the influence that cytokine–cytokine cross talk has on promoting membrane release of TNF-family ligands. Barker and colleagues recently demonstrated that pro-NGF engagement of p75-NGF receptor induces apoptosis in retinal ganglion cells by inducing the expression of TNFα (Lebrun-Julien et al. 2010). Taken together with the finding that p75-NGF receptor can also induce transcriptional upregulation of ADAM17 in sympathetic neurons, it suggests that p75-NGF receptor controls not only the expression of TNFα but perhaps also its release (Kenchappa et al. 2010). A provocative-yet-untested implication of this finding is that TNF-family signaling may control the processing of its own ligand. There is clear precedent for such a positive-feedback mechanism in the EGF-family ligand, TGFα, which is shed proteolytically after RTK signaling (Fan and Derynck 1999). While there are many elegant models for processing of RTK ligands (Shvartsman et al. 2002a, b), similar approaches have not yet been applied to TNF-family ligands for analyzing their potential to create autocrine circuits.

8.7 Conclusions and Future Directions

There would be great benefit in gaining a systems-level understanding of cytokine–cytokine cross talk, even if all the molecular details have not been fully unraveled. Systems models could evaluate reported mechanisms of cross talk and distinguish those that are quantitatively important for coupling two cytokine pathways. Artificially disrupting critical cross talk nodes would allow us to "untangle the wires" and determine the physiological importance of cross talk during development and disease. More creatively, one could engineer synthetic signaling nodes, which create cross talk between cytokines that would not ordinarily exist (Howard et al. 2003; Jay et al. 2011; Yeh et al. 2007). Indeed, there is precedent for de novo cross talk during the evolution of human cancers (Engelman et al. 2007; Johannessen et al. 2010; Nazarian et al. 2010; Sergina et al. 2007). Likewise, in neurodegenerative disorders, developmental signaling pathways are inappropriately reengaged leading to novel and ultimately destructive signaling nodes (Nikolaev et al. 2009). A similar phenomenon can be observed in spinal cord injury, where increased TNF-family signaling prevents axon regrowth and promotes death of neurons (Beattie et al. 2002; Shao et al. 2005). The ability to engineer cross talk between cytokine pathways promises to yield novel therapies for a wide array of historically intractable diseases.

References

Akca H, Akan SY, Yanikoglu A, Ozes ON (2003) Suppression of TNF-α mediated apoptosis by EGF in TNF-α sensitive human cervical carcinoma cell line. Growth Factors 21:31–39

Albeck JG, MacBeath G, White FM, Sorger PK, Lauffenburger DA, Gaudet S (2006) Collecting and organizing systematic sets of protein data. Nat Rev Mol Cell Biol 7:803–812

Aldridge BB, Burke JM, Lauffenburger DA, Sorger PK (2006) Physicochemical modelling of cell signalling pathways. Nat Cell Biol 8:1195–1203

Aldridge BB, Saez-Rodriguez J, Muhlich JL, Sorger PK, Lauffenburger DA (2009) Fuzzy logic analysis of kinase pathway crosstalk in TNF/EGF/insulin-induced signaling. PLoS Comput Biol 5:e1000340

Alexopoulou L, Kranidioti K, Xanthoulea S, Denis M, Kotanidou A, Douni E, Blackshear PJ, Kontoyiannis DL, Kollias G (2006) Transmembrane TNF protects mutant mice against intracellular bacterial infections, chronic inflammation and autoimmunity. Eur J Immunol 36:2768–2780

Allan LA, Morrice N, Brady S, Magee G, Pathak S, Clarke PR (2003) Inhibition of caspase-9 through phosphorylation at Thr 125 by ERK MAPK. Nat Cell Biol 5:647–654

Altieri DC (2003) Validating survivin as a cancer therapeutic target. Nat Rev Cancer 3:46–54

Amit I, Citri A, Shay T, Lu Y, Katz M, Zhang F, Tarcic G, Siwak D, Lahad J, Jacob-Hirsch J et al (2007) A module of negative feedback regulators defines growth factor signaling. Nat Genet 39:503–512

Ashkenazi A, Dixit VM (1998) Death receptors: signaling and modulation. Science 281:1305–1308

Avraham R, Yarden Y (2011) Feedback regulation of EGFR signalling: decision making by early and delayed loops. Nat Rev Mol Cell Biol 12:104–117

Bamji SX, Majdan M, Pozniak CD, Belliveau D, Aloyz R, Kohn J, Causing CG, Miller FD (1998) The p75 neurotrophin receptor mediates neuronal apoptosis and is essential for naturally occurring sympathetic neuron death. J Cell Biol 140:911–923

Bange J, Zwick E, Ullrich A (2001) Molecular targets for breast cancer therapy and prevention. Nat Med 7:548–552

Beattie MS, Harrington AW, Lee R, Kim JY, Boyce SL, Longo FM, Bresnahan JC, Hempstead BL, Yoon SO (2002) ProNGF induces p75-mediated death of oligodendrocytes following spinal cord injury. Neuron 36:375–386

Bedard S, Marcotte B, Marette A (1998) Insulin inhibits inducible nitric oxide synthase in skeletal muscle cells. Diabetologia 41:1523–1527

Bhalla US, Iyengar R (1999) Emergent properties of networks of biological signaling pathways. Science 283:381–387

Blume-Jensen P, Hunter T (2001) Oncogenic kinase signalling. Nature 411:355–365

Borisov N, Aksamitiene E, Kiyatkin A, Legewie S, Berkhout J, Maiwald T, Kaimachnikov NP, Timmer J, Hoek JB, Kholodenko BN (2009) Systems-level interactions between insulin-EGF networks amplify mitogenic signaling. Mol Syst Biol 5:256

Brunet A, Bonni A, Zigmond MJ, Lin MZ, Juo P, Hu LS, Anderson MJ, Arden KC, Blenis J, Greenberg ME (1999) Akt promotes cell survival by phosphorylating and inhibiting a Forkhead transcription factor. Cell 96:857–868

Canault M, Peiretti F, Mueller C, Kopp F, Morange P, Rihs S, Portugal H, Juhan-Vague I, Nalbone G (2004) Exclusive expression of transmembrane TNF-α in mice reduces the inflammatory response in early lipid lesions of aortic sinus. Atherosclerosis 172:211–218

Cantley LC, Neel BG (1999) New insights into tumor suppression: PTEN suppresses tumor formation by restraining the phosphoinositide 3-kinase/AKT pathway. Proc Natl Acad Sci USA 96:4240–4245

Cardone MH, Roy N, Stennicke HR, Salvesen GS, Franke TF, Stanbridge E, Frisch S, Reed JC (1998) Regulation of cell death protease caspase-9 by phosphorylation. Science 282:1318–1321

Chailler P, Menard D (1999) Ontogeny of EGF receptors in the human gut. Front Biosci 4:D87–101

Chatterjee MS, Purvis JE, Brass LF, Diamond SL (2010) Pairwise agonist scanning predicts cellular signaling responses to combinatorial stimuli. Nat Biotechnol 28:727–732

Chen WN, Woodbury RL, Kathmann LE, Opresko LK, Zangar RC, Wiley HS, Thrall BD (2004) Induced autocrine signaling through the epidermal growth factor receptor contributes to the response of mammary epithelial cells to tumor necrosis factor alpha. J Biol Chem 279:18488–18496

Cheng CS, Feldman KE, Lee J, Verma S, Huang DB, Huynh K, Chang M, Ponomarenko JV, Sun SC, Benedict CA et al (2011) The specificity of innate immune responses is enforced by repression of interferon response elements by NF-κB p50. Sci Signal 4:ra11

Cohen-Saidon C, Cohen AA, Sigal A, Liron Y, Alon U (2009) Dynamics and variability of ERK2 response to EGF in individual living cells. Mol Cell 36:885–893

Cosgrove BD, Cheng C, Pritchard JR, Stolz DB, Lauffenburger DA, Griffith LG (2008) An inducible autocrine cascade regulates rat hepatocyte proliferation and apoptosis responses to tumor necrosis factor-alpha. Hepatology 48:276–288

Covert MW, Leung TH, Gaston JE, Baltimore D (2005) Achieving stability of lipopolysaccharide-induced NF-κB activation. Science 309:1854–1857

Danial NN, Gramm CF, Scorrano L, Zhang CY, Krauss S, Ranger AM, Datta SR, Greenberg ME, Licklider LJ, Lowell BB et al (2003) BAD and glucokinase reside in a mitochondrial complex that integrates glycolysis and apoptosis. Nature 424:952–956

Datta SR, Dudek H, Tao X, Masters S, Fu H, Gotoh Y, Greenberg ME (1997) Akt phosphorylation of BAD couples survival signals to the cell-intrinsic death machinery. Cell 91:231–241

Datta SR, Brunet A, Greenberg ME (1999) Cellular survival: a play in three Akts. Genes Dev 13:2905–2927

Datta SR, Ranger AM, Lin MZ, Sturgill JF, Ma YC, Cowan CW, Dikkes P, Korsmeyer SJ, Greenberg ME (2002) Survival factor-mediated BAD phosphorylation raises the mitochondrial threshold for apoptosis. Dev Cell 3:631–643

Delhase M, Li N, Karin M (2000) Kinase regulation in inflammatory response. Nature 406:367–368

Deng J, Miller SA, Wang HY, Xia W, Wen Y, Zhou BP, Li Y, Lin SY, Hung MC (2002) β-Catenin interacts with and inhibits NF-κ B in human colon and breast cancer. Cancer Cell 2:323–334

Deppmann CD, Mihalas S, Sharma N, Lonze BE, Niebur E, Ginty DD (2008) A model for neuronal competition during development. Science 320:369–373
DeWitt AE, Dong JY, Wiley HS, Lauffenburger DA (2001) Quantitative analysis of the EGF receptor autocrine system reveals cryptic regulation of cell response by ligand capture. J Cell Sci 114:2301–2313
Downward J (2001) The ins and outs of signalling. Nature 411:759–762
Engelman JA, Zejnullahu K, Mitsudomi T, Song Y, Hyland C, Park JO, Lindeman N, Gale CM, Zhao X, Christensen J et al (2007) MET amplification leads to gefitinib resistance in lung cancer by activating ERBB3 signaling. Science 316:1039–1043
Fan H, Derynck R (1999) Ectodomain shedding of TGF-α and other transmembrane proteins is induced by receptor tyrosine kinase activation and MAP kinase signaling cascades. EMBO J 18:6962–6972
Fernandez-Marcos PJ, Abu-Baker S, Joshi J, Galvez A, Castilla EA, Canamero M, Collado M, Saez C, Moreno-Bueno G, Palacios J et al (2009) Simultaneous inactivation of Par-4 and PTEN in vivo leads to synergistic NF-κB activation and invasive prostate carcinoma. Proc Natl Acad Sci USA 106:12962–12967
Franke TF, Kaplan DR, Cantley LC (1997) PI3K: downstream AKTion blocks apoptosis. Cell 88:435–437
Garcia-Lloret MI, Yui J, Winkler-Lowen B, Guilbert LJ (1996) Epidermal growth factor inhibits cytokine-induced apoptosis of primary human trophoblasts. J Cell Physiol 167:324–332
Garmaroudi FS, Marchant D, Si X, Khalili A, Bashashati A, Wong BW, Tabet A, Ng RT, Murphy K, Luo H et al (2010) Pairwise network mechanisms in the host signaling response to coxsackie virus B3 infection. Proc Natl Acad Sci USA 107:17053–17058
Gaudet S, Janes KA, Albeck JG, Pace EA, Lauffenburger DA, Sorger PK (2005) A compendium of signals and responses triggered by prodeath and prosurvival cytokines. Mol Cell Proteomics 4:1569–1590
Gibson S, Tu S, Oyer R, Anderson SM, Johnson GL (1999) Epidermal growth factor protects epithelial cells against Fas-induced apoptosis requirement for Akt activation. J Biol Chem 274:17612–17618
Goetze S, Blaschke F, Stawowy P, Bruemmer D, Spencer C, Graf K, Grafe M, Law RE, Fleck E (2001) TNFalpha inhibits insulin's antiapoptotic signaling in vascular smooth muscle cells. Biochem Biophys Res Commun 287:662–670
Gordus A, Krall JA, Beyer EM, Kaushansky A, Wolf-Yadlin A, Sevecka M, Chang BH, Rush J, MacBeath G (2009) Linear combinations of docking affinities explain quantitative differences in RTK signaling. Mol Syst Biol 5:235
Graham NA, Asthagiri AR (2004) Epidermal growth factor-mediated T-cell factor/lymphoid enhancer factor transcriptional activity is essential but not sufficient for cell cycle progression in nontransformed mammary epithelial cells. J Biol Chem 279:23517–23524
Gustin JA, Ozes ON, Akca H, Pincheira R, Mayo LD, Li Q, Guzman JR, Korgaonkar CK, Donner DB (2004) Cell type-specific expression of the IkappaB kinases determines the significance of phosphatidylinositol 3-kinase/Akt signaling to NF-κ B activation. J Biol Chem 279:1615–1620
Hanahan D, Weinberg RA (2000) The hallmarks of cancer. Cell 100:57–70
Hanahan D, Weinberg RA (2011) Hallmarks of cancer: the next generation. Cell 144:646–674
Holtmann MH, Neurath MF (2004) Differential TNF-signaling in chronic inflammatory disorders. Curr Mol Med 4:439–444
Howard PL, Chia MC, Del Rizzo S, Liu FF, Pawson T (2003) Redirecting tyrosine kinase signaling to an apoptotic caspase pathway through chimeric adaptor proteins. Proc Natl Acad Sci USA 100:11267–11272
Hsueh RC, Natarajan M, Fraser I, Pond B, Liu J, Mumby S, Han H, Jiang LI, Simon MI, Taussig R et al (2009) Deciphering signaling outcomes from a system of complex networks. Sci Signal 2:ra22
Huse M, Kuriyan J (2002) The conformational plasticity of protein kinases. Cell 109:275–282

Idriss HT, Naismith JH (2000) TNF alpha and the TNF receptor superfamily: structure-function relationship(s). Microsc Res Tech 50:184–195
Janes KA (2010) Paring down signaling complexity. Nat Biotechnol 28:681–682
Janes KA, Lauffenburger DA (2006) A biological approach to computational models of proteomic networks. Curr Opin Chem Biol 10:73–80
Janes KA, Yaffe MB (2006) Data-driven modelling of signal-transduction networks. Nat Rev Mol Cell Biol 7:820–828
Janes KA, Albeck JG, Gaudet S, Sorger PK, Lauffenburger DA, Yaffe MB (2005) A systems model of signaling identifies a molecular basis set for cytokine-induced apoptosis. Science 310:1646–1653
Janes KA, Gaudet S, Albeck JG, Nielsen UB, Lauffenburger DA, Sorger PK (2006) The response of human epithelial cells to TNF involves an inducible autocrine cascade. Cell 124:1225–1239
Janes KA, Reinhardt HC, Yaffe MB (2008) Cytokine-induced signaling networks prioritize dynamic range over signal strength. Cell 135:343–354
Janeway CA, Travers P, Walport M, Shlomchik MJ (2001) Immunobiology, 5th edn. Garland, New York
Jay SM, Kurtagic E, Alvarez LM, de Picciotto S, Sanchez E, Hawkins JF, Prince RN, Guerrero Y, Treasure CL, Lee RT et al (2011) Engineered bivalent ligands to bias ErbB receptor-mediated signaling and phenotypes. J Biol Chem 286:27729–40
Johannessen CM, Boehm JS, Kim SY, Thomas SR, Wardwell L, Johnson LA, Emery CM, Stransky N, Cogdill AP, Barretina J et al (2010) COT drives resistance to RAF inhibition through MAP kinase pathway reactivation. Nature 468:968–972
Karin M, Ben-Neriah Y (2000) Phosphorylation meets ubiquitination: the control of NF-[κ]B activity. Annu Rev Immunol 18:621–663
Kasibhatla S, Brunner T, Genestier L, Echeverri F, Mahboubi A, Green DR (1998) DNA damaging agents induce expression of Fas ligand and subsequent apoptosis in T lymphocytes via the activation of NF-κ B and AP-1. Mol Cell 1:543–551
Kenchappa RS, Tep C, Korade Z, Urra S, Bronfman FC, Yoon SO, Carter BD (2010) p75 Neurotrophin receptor-mediated apoptosis in sympathetic neurons involves a biphasic activation of JNK and up-regulation of tumor necrosis factor-alpha-converting enzyme/ADAM17. J Biol Chem 285:20358–20368
Kim S, Domon-Dell C, Kang J, Chung DH, Freund JN, Evers BM (2004) Down-regulation of the tumor suppressor PTEN by the tumor necrosis factor-alpha/nuclear factor-κB (NF-κB)-inducing kinase/NF-κB pathway is linked to a default IκB-α autoregulatory loop. J Biol Chem 279:4285–4291
Kim T, Yoon J, Cho H, Lee WB, Kim J, Song YH, Kim SN, Yoon JH, Kim-Ha J, Kim YJ (2005) Downregulation of lipopolysaccharide response in Drosophila by negative crosstalk between the AP1 and NF-κB signaling modules. Nat Immunol 6:211–218
Kohn J, Aloyz RS, Toma JG, Haak-Frendscho M, Miller FD (1999) Functionally antagonistic interactions between the TrkA and p75 neurotrophin receptors regulate sympathetic neuron growth and target innervation. J Neurosci 19:5393–5408
Kumar D, Srikanth R, Ahlfors H, Lahesmaa R, Rao KV (2007) Capturing cell-fate decisions from the molecular signatures of a receptor-dependent signaling response. Mol Syst Biol 3:150
LaCasse EC, Baird S, Korneluk RG, MacKenzie AE (1998) The inhibitors of apoptosis (IAPs) and their emerging role in cancer. Oncogene 17:3247–3259
Lauffenburger DA, Forsten KE, Will B, Wiley HS (1995) Molecular/cell engineering approach to autocrine ligand control of cell function. Ann Biomed Eng 23:208–215
Lavrik I, Golks A, Krammer PH (2005) Death receptor signaling. J Cell Sci 118:265–267
Lebrun-Julien F, Bertrand MJ, De Backer O, Stellwagen D, Morales CR, Di Polo A, Barker PA (2010) ProNGF induces TNFalpha-dependent death of retinal ganglion cells through a p75NTR non-cell-autonomous signaling pathway. Proc Natl Acad Sci USA 107:3817–3822
Lemmon MA, Schlessinger J (2010) Cell signaling by receptor tyrosine kinases. Cell 141:1117–1134

Li ZW, Chu W, Hu Y, Delhase M, Deerinck T, Ellisman M, Johnson R, Karin M (1999) The IKKbeta subunit of IkappaB kinase (IKK) is essential for nuclear factor kappaB activation and prevention of apoptosis. J Exp Med 189:1839–1845

Locksley RM, Killeen N, Lenardo MJ (2001) The TNF and TNF receptor superfamilies: integrating mammalian biology. Cell 104:487–501

Majdan M, Walsh GS, Aloyz R, Miller FD (2001) TrkA mediates developmental sympathetic neuron survival in vivo by silencing an ongoing p75NTR-mediated death signal. J Cell Biol 155:1275–1285

Manning G, Whyte DB, Martinez R, Hunter T, Sudarsanam S (2002) The protein kinase complement of the human genome. Science 298:1912–1934

McKay MM, Morrison DK (2007) Integrating signals from RTKs to ERK/MAPK. Oncogene 26:3113–3121

Miller-Jensen K, Janes KA, Brugge JS, Lauffenburger DA (2007) Common effector processing mediates cell-specific responses to stimuli. Nature 448:604–608

Modur V, Nagarajan R, Evers BM, Milbrandt J (2002) FOXO proteins regulate tumor necrosis factor-related apoptosis inducing ligand expression Implications for PTEN mutation in prostate cancer. J Biol Chem 277:47928–47937

Monine MI, Berezhkovskii AM, Joslin EJ, Wiley HS, Lauffenburger DA, Shvartsman SY (2005) Ligand accumulation in autocrine cell cultures. Biophys J 88:2384–2390

Mueller C, Corazza N, Trachsel-Loseth S, Eugster HP, Buhler-Jungo M, Brunner T, Imboden MA (1999) Noncleavable transmembrane mouse tumor necrosis factor-alpha (TNFalpha) mediates effects distinct from those of wild-type TNFalpha in vitro and in vivo. J Biol Chem 274:38112–38118

Natarajan M, Lin KM, Hsueh RC, Sternweis PC, Ranganathan R (2006) A global analysis of cross-talk in a mammalian cellular signalling network. Nat Cell Biol 8:571–580

Nazarian R, Shi H, Wang Q, Kong X, Koya RC, Lee H, Chen Z, Lee MK, Attar N, Sazegar H et al (2010) Melanomas acquire resistance to B-RAF(V600E) inhibition by RTK or N-RAS upregulation. Nature 468:973–977

Nikolaev A, McLaughlin T, O'Leary DD, Tessier-Lavigne M (2009) APP binds DR6 to trigger axon pruning and neuron death via distinct caspases. Nature 457:981–989

Obata T, Yaffe MB, Leparc GG, Piro ET, Maegawa H, Kashiwagi A, Kikkawa R, Cantley LC (2000) Peptide and protein library screening defines optimal substrate motifs for AKT/PKB. J Biol Chem 275:36108–36115

Ozes ON, Mayo LD, Gustin JA, Pfeffer SR, Pfeffer LM, Donner DB (1999) NF-κB activation by tumour necrosis factor requires the Akt serine-threonine kinase. Nature 401:82–85

Ozoren N, El-Deiry WS (2003) Cell surface death receptor signaling in normal and cancer cells. Semin Cancer Biol 13:135–147

Pawson T, Scott JD (1997) Signaling through scaffold, anchoring, and adaptor proteins. Science 278:2075–2080

Perkins ND (2004) NF-κB: tumor promoter or suppressor? Trends Cell Biol 14:64–69

Perry RT, Collins JS, Wiener H, Acton R, Go RC (2001) The role of TNF and its receptors in Alzheimer's disease. Neurobiol Aging 22:873–883

Pitti RM, Marsters SA, Lawrence DA, Roy M, Kischkel FC, Dowd P, Huang A, Donahue CJ, Sherwood SW, Baldwin DT et al (1998) Genomic amplification of a decoy receptor for Fas ligand in lung and colon cancer. Nature 396:699–703

Qian H, Hausman DB, Compton MM, Martin RJ, Della-Fera MA, Hartzell DL, Baile CA (2001) TNFalpha induces and insulin inhibits caspase 3-dependent adipocyte apoptosis. Biochem Biophys Res Commun 284:1176–1183

Rodriguez J, Chen HH, Lin SC, Lazebnik Y (2000) Caspase phosphorylation, cell death, and species variability. Science 287:1363

Rutgeerts P, Van Assche G, Vermeire S (2004) Optimizing anti-TNF treatment in inflammatory bowel disease. Gastroenterology 126:1593–1610

Ruuls SR, Hoek RM, Ngo VN, McNeil T, Lucian LA, Janatpour MJ, Korner H, Scheerens H, Hessel EM, Cyster JG et al (2001) Membrane-bound TNF supports secondary lymphoid organ structure but is subservient to secreted TNF in driving autoimmune inflammation. Immunity 15:533–543

Saez-Rodriguez J, Alexopoulos LG, Epperlein J, Samaga R, Lauffenburger DA, Klamt S, Sorger PK (2009) Discrete logic modelling as a means to link protein signalling networks with functional analysis of mammalian signal transduction. Mol Syst Biol 5:331

Scaffidi C, Fulda S, Srinivasan A, Friesen C, Li F, Tomaselli KJ, Debatin KM, Krammer PH, Peter ME (1998) Two CD95 (APO-1/Fas) signaling pathways. EMBO J 17:1675–1687

Schlessinger J (2000) Cell signaling by receptor tyrosine kinases. Cell 103:211–225

Sergina NV, Rausch M, Wang D, Blair J, Hann B, Shokat KM, Moasser MM (2007) Escape from HER-family tyrosine kinase inhibitor therapy by the kinase-inactive HER3. Nature 445:437–441

Shao Z, Browning JL, Lee X, Scott ML, Shulga-Morskaya S, Allaire N, Thill G, Levesque M, Sah D, McCoy JM et al (2005) TAJ/TROY, an orphan TNF receptor family member, binds Nogo-66 receptor 1 and regulates axonal regeneration. Neuron 45:353–359

Sheikh MS, Huang Y, Fernandez-Salas EA, El-Deiry WS, Friess H, Amundson S, Yin J, Meltzer SJ, Holbrook NJ, Fornace AJ Jr (1999) The antiapoptotic decoy receptor TRID/TRAIL-R3 is a p53-regulated DNA damage-inducible gene that is overexpressed in primary tumors of the gastrointestinal tract. Oncogene 18:4153–4159

Sheng M, Greenberg ME (1990) The regulation and function of c-fos and other immediate early genes in the nervous system. Neuron 4:477–485

Shvartsman SY, Hagan MP, Yacoub A, Dent P, Wiley HS, Lauffenburger DA (2002a) Autocrine loops with positive feedback enable context-dependent cell signaling. Am J Physiol Cell Physiol 282:C545–559

Shvartsman SY, Muratov CB, Lauffenburger DA (2002b) Modeling and computational analysis of EGF receptor-mediated cell communication in Drosophila oogenesis. Development 129:2577–2589

Singh P, Rubin N (1993) Insulin-like growth factors and binding proteins in colon cancer. Gastroenterology 105:1218–1237

Song W, Volosin M, Cragnolini AB, Hempstead BL, Friedman WJ (2010) ProNGF induces PTEN via p75NTR to suppress Trk-mediated survival signaling in brain neurons. J Neurosci 30:15608–15615

Ullrich A, Schlessinger J (1990) Signal transduction by receptors with tyrosine kinase activity. Cell 61:203–212

Vilar M, Charalampopoulos I, Kenchappa RS, Simi A, Karaca E, Reversi A, Choi S, Bothwell M, Mingarro I, Friedman WJ et al (2009) Activation of the p75 neurotrophin receptor through conformational rearrangement of disulphide-linked receptor dimers. Neuron 62:72–83

Wang Z, Cao N, Nantajit D, Fan M, Liu Y, Li JJ (2008) Mitogen-activated protein kinase phosphatase-1 represses c-Jun NH2-terminal kinase-mediated apoptosis via NF-κB regulation. J Biol Chem 283:21011–21023

Ware CF (2005) Network communications: lymphotoxins, LIGHT, and TNF. Annu Rev Immunol 23:787–819

Wehrman T, He X, Raab B, Dukipatti A, Blau H, Garcia KC (2007) Structural and mechanistic insights into nerve growth factor interactions with the TrkA and p75 receptors. Neuron 53:25–38

Weiner FR, Smith PJ, Wertheimer S, Rubin CS (1991) Regulation of gene expression by insulin and tumor necrosis factor alpha in 3T3-L1 cells. Modulation of the transcription of genes encoding acyl-CoA synthetase and stearoyl-CoA desaturase-1. J Biol Chem 266:23525–23528

Werner SL, Barken D, Hoffmann A (2005) Stimulus specificity of gene expression programs determined by temporal control of IKK activity. Science 309:1857–1861

Wu Y, Tewari M, Cui S, Rubin R (1996) Activation of the insulin-like growth factor-I receptor inhibits tumor necrosis factor-induced cell death. J Cell Physiol 168:499–509

Wu JJ, Roth RJ, Anderson EJ, Hong EG, Lee MK, Choi CS, Neufer PD, Shulman GI, Kim JK, Bennett AM (2006) Mice lacking MAP kinase phosphatase-1 have enhanced MAP kinase activity and resistance to diet-induced obesity. Cell Metab 4:61–73

Yeh BJ, Rutigliano RJ, Deb A, Bar-Sagi D, Lim WA (2007) Rewiring cellular morphology pathways with synthetic guanine nucleotide exchange factors. Nature 447:596–600

Yuzawa S, Opatowsky Y, Zhang Z, Mandiyan V, Lax I, Schlessinger J (2007) Structural basis for activation of the receptor tyrosine kinase KIT by stem cell factor. Cell 130:323–334

Zhang R, Shah MV, Yang J, Nyland SB, Liu X, Yun JK, Albert R, Loughran TP Jr (2008) Network model of survival signaling in large granular lymphocyte leukemia. Proc Natl Acad Sci USA 105:16308–16313

Chapter 9
Genetic and Genomic Dissection of Apoptosis Signaling

Christina Falschlehner and Michael Boutros

Abstract Systematic loss-of-function approaches have significantly contributed to our understanding of apoptotic signaling networks. In this book chapter, we will review classical forward genetic approaches and high-throughput RNA interference screens that led to the identification of key factors regulating cellular survival and cell death. We will describe how synthetic lethal screens helped to dissect regulatory networks and contributed to the development of targeted drugs. We will further provide an outlook on future directions of this area in systems biology.

9.1 Genetic Screens for Apoptosis

The search for modulators of cell survival began in the 1960s, right after the discovery of programmed cell death in insects (Lockshin and Williams 1965) and further studies in the nematode *Caenorhabditis elegans* (*C. elegans*) (Sulston and Horvitz 1977). Using classical forward genetic approaches, several genes were identified that play important roles in apoptosis induction and inhibition. *C. elegans* proved to be a great model organism to study cell death as the developmental fate of each somatic cell is predefined. The adult *C. elegans* hermaphrodite contains exactly 959 somatic cell nuclei as precisely 131 of 1,090 cells undergo apoptosis in a time-controlled manner during embryogenesis (Sulston and Horvitz 1977).

Sydney Brenner was the first to introduce the worm *C. elegans* to the scientific community (Brenner 1974). In 1974, he showed that point mutations of *C. elegans* strains can be easily obtained by the use of mutagens such as ethyl methane sulphonate (EMS) (Epstein et al. 1974). In the following years, a plethora of mutant

C. Falschlehner • M. Boutros (✉)
German Cancer Research Center, Division of Signaling and Functional Genomics, and Heidelberg University, Department of Cell and Molecular Biology, Medical Faculty Mannheim, Im Neuenheimer Feld 580, 69120 Heidelberg, Germany
e-mail: c.falschlehner@dkfz.de; m.boutros@dkfz.de

I.N. Lavrik (ed.), *Systems Biology of Apoptosis*, DOI 10.1007/978-1-4614-4009-3_9,

strains were described. The analysis of mutants with similar phenotypes revealed that the affected genes were often part of the same signaling pathway. For example, many genes of the respective mutant strains with defects in vulva differentiation could be mapped to two pathways: the Notch and RAS pathway, both required for epidermal differentiation (Ferguson et al. 1987; Wang and Sternberg 2001). The vulva is required for egg laying, so the mutants could easily be scored by the fact that no eggs were present. Some of these mutants later proved to be very useful for the dissection of the apoptotic signaling pathway in *C. elegans*. The first experiments on programmed cell death in *C. elegans* were undertaken by Robert Horvitz and colleagues (1994). They used Nomarski optics, a microscopy technique that shows high contrast differences in transparent samples, to visualize all cells of the worm. Using this technique, mutants that showed an increased or decreased cell count could be scored. The observed change in cell number was either due to a defect in proliferation or a defect to undergo programmed cell death. Looking at diverse mutant strains, it became clear that programmed cell death is predetermined as exactly the same 131 cells always died during embryogenesis and were missing in the adult worm.

The first gene identified that alters cell numbers in *C. elegans* was nuc-1 which was found to play a role in DNA degradation (Sulston and Horvitz 1981; Wu et al. 2000b). In the following years, mutants of ced-1 (cell death abnormality 1) and ced-2, both showing defects in autophagy were described (Ellis et al. 1991). Under normal circumstances, apoptotic cells are rapidly engulfed by phagocytic cells and are thus difficult to visualize. Fortunately, the use of ced-1 mutants allowed easy imaging of apoptotic cells as they remain visible for a longer period of time. Thus, genes affecting apoptotic cell death could be identified. This approach led to the discovery of ced-3, the *C. elegans* homologue of human caspases (Schwartz and Osborne 1994; Yuan and Horvitz 1990). As mentioned earlier, mutants defective in egg laying that were previously generated, turned out to be meaningful for the dissection of the cell death pathway in *C. elegans*. A mutant, deficient in egg laying named egl-1 (egg laying defective) did not contain hermaphrodite specific motor neurons (HSNs) which are required for egg laying (Trent et al. 1983). In males that do not have a vulva for egg laying and thus do not contain HSNs, it was shown that these neurons undergo programmed cell death during development. Introduction of the proapoptotic protease ced-3 into the egl-1 mutant restored egg laying in these mutants and thus confirmed the essential role of ced-3 for apoptosis (Ellis and Horvitz 1986). Egl-1 was later identified as a member of the BH3-domain only Bcl-2 (B-cell lymphoma 2) family, similar to human Bid (BH3 interacting domain death agonist) or Bim (BCL2-like 11) (Huang and Strasser 2000). A few years later, ced-4, the *C. elegans* homologue of human Apaf-1 (Apoptotic protease-activating factor 1) and ced-9, the homologue of human Bcl-2 were identified (Cecconi 1999; Ellis and Horvitz 1986; Hengartner et al. 1992; Hengartner and Horvitz 1994; Yuan and Horvitz 1990).

Classical forward genetic approaches were also pursued in other model organisms such as the fly *Drosophila melanogaster* or the zebrafish *Danio Rerio* (Patton and Zon 2001; St Johnston 2002). It became quickly clear that the apoptotic machinery in flies and vertebrates is more complex and requires additional regulatory factors. An important gene family that has been identified is the inhibitor of apoptosis (IAP) family. In contrast to *C. elegans*, caspases in flies and vertebrates have to be kept in check by IAPs to avoid uncontrolled substrate cleavage (Salvesen and Duckett 2002; Vaux and Silke 2005). The first inhibitor of apoptosis protein in *Drosophila*, DIAP1, was identified by a genetic screen for suppressors of reaper-dependent cell death in 1995 (Hay et al. 1995). The E3 ubiquitin ligase DIAP1 was shown to be essential for the survival of the flies. Homozygous flies lacking DIAP1 died due to massive caspase-dependent cell death (Rodriguez et al. 2002). In flies, the homologue of human caspase-9 named Dronc is constitutively activated by Ark (human Apaf-1) and executes the cell death program, unless it is bound to and inhibited by DIAP1 (Hay et al. 1995; Lisi et al. 2000; Wang et al. 1999). Various substrates of Dronc were subsequently identified including the effector caspases DRICE and Dcp-1 as well as DIAP1 (Kumar and Doumanis 2000). The previously mentioned protein reaper and the proteins grim and hid were discovered as key regulators of apoptosis during *Drosophila* embryogenesis (Abrams et al. 1993; White et al. 1994). Mutant flies that lacked reaper had more cells and were not able to hatch (White et al. 1994). Reaper, grim, and hid were later shown to share significant similarity with the human IAP-antagonists Smac/Diablo and Omi/HtrA2 (Chai et al. 2000; Liu et al. 2000; Martins et al. 2002; Srinivasula et al. 2000; Wu et al. 2000a). All three proteins can bind with high affinity to DIAP1 preventing it to interact with and inhibit the initiator caspase Dronc. As a result, fully activated Dronc drives destruction of the cell finally leading to cell death (Igaki et al. 2002).

9.2 Genetic Screens by RNA Interference

9.2.1 Introduction to RNAi

RNA interference (RNAi) is a sequence-specific, posttranscriptional silencing mechanism that is mediated by double-stranded RNA (dsRNA) molecules. RNAi occurs when dsRNA enters the cell and triggers degradation of mRNA molecules that carry complementary sequences.

The mechanism of RNAi was first discovered in *C. elegans*. When long dsRNAs were injected into these roundworms, they blocked the expression of endogenous genes that shared the same sequence (Fire et al. 1998; Tabara et al. 1998). It was initially thought that RNAi has evolved as a defense mechanism against viruses that carry dsRNA. Today, we know that the RNAi machinery is also used to regulate endogenous gene activity by so-called microRNAs (miRNAs). miRNAs can be found in almost all higher eukaryotes and are often located in intergenic regions of chromosomal DNA (Bentwich et al. 2005; Lau et al. 2001). Upon transcriptional activation, endogenous miRNAs are first produced as approximately 70 nucleotide-long pri-miRNAs by RNA polymerase II (Han et al. 2004). These hairpin-shaped pri-miRNAs are further processed by the nuclease Drosha into pre-miRNAs that are transported from the nucleus into the cytoplasm via exportin 5. In the cytoplasm the RNA hairpin is recognized by the RNAse III endonuclease Dicer and cleaved into mature 19–25 nucleotide-long miRNA duplex structures (Preall and Sontheimer 2005). These mature miRNAs are incoporated into the RNA-induced silencing complex (RISC) where they are unwound and paired to complementary mRNA, preferentially to the 3′ untranslated region (UTR) of the transcript. Exact base-pairing results in cleavage and degradation of mRNA while imperfect matching leads to translational repression (Good 2003; Meister et al. 2004).

A similar mechanism is triggered when dsRNAs are introduced into the cell. They are recognized and cleaved by Dicer into small interfering RNAs (siRNAs). siRNAs usually have a length of 21–23 base pairs with a two nucleotide-long overhang on their 3′ ends. Chemically synthesized siRNAs mimic Dicer-processed molecules. siRNAs are incorporated into the RISC complex where they regulate target gene expression similar to the described mechanism above. However, in contrast to miRNAs, no preferential binding to UTRs of mRNA transcripts has been observed. Most artificially designed siRNAs are designed to target the open reading frame of the respective gene.

Dicer-processed siRNAs that share complementary regions with promoters might translocate back to the nucleus and cause transcriptional gene silencing (Mette et al. 2000; Zeng and Cullen 2002). This is probably achieved by DNA or histone methylation leading to tightly packed heterochromatin. However, the exact molecular mechanism of Dicer-induced gene silencing in the nucleus is currently not fully understood. A simplified representation of the RNAi mechanism is shown in Fig. 9.1.

With the availability of whole genome sequences, directed gene silencing by RNAi has complemented classical forward genetic screening approaches. Mapping of genetic point mutations that caused a certain phenotype is often time-consuming and difficult to perform for many identified alleles. In contrast, the use of RNAi libraries allowed the generation of genome-wide loss-of-function data sets in a relatively short time and enabled reconstruction of complex signaling networks and modeling of biological processes. In the past few years, RNAi libraries that target almost all annotated genes in plants, worms, flies, mice, and humans have been created and are widely used in high-throughput screening approaches (Boutros and Ahringer 2008; Hu et al. 2010; Karlas et al. 2010; Neely et al. 2010; Zhang et al. 2009). Depending on the model organism, RNAi libraries are available in various formats. In worm, fly, or plant models, long dsRNAs (100–700 bp) can be used to induce gene silencing while mammalian cells have to be treated with shorter siRNAs (21–23 bp) to avoid an interferon response (Elbashir et al. 2002).

DsRNA can be introduced to the cells by various means *in vitro* and *in vivo*. Common delivery methods for cell culture-based siRNA screens include liposomal transfection and electroporation. Chemically synthesized siRNAs can be modified to increase their stability and efficacy. In most cases the ribose at the 2′ position is methylated (2′Ome), allylated (2′Oal), or modified nucleotides such as locked nucleic acids (LNAs) or phosphorothioate interlinked nucleotides are used to

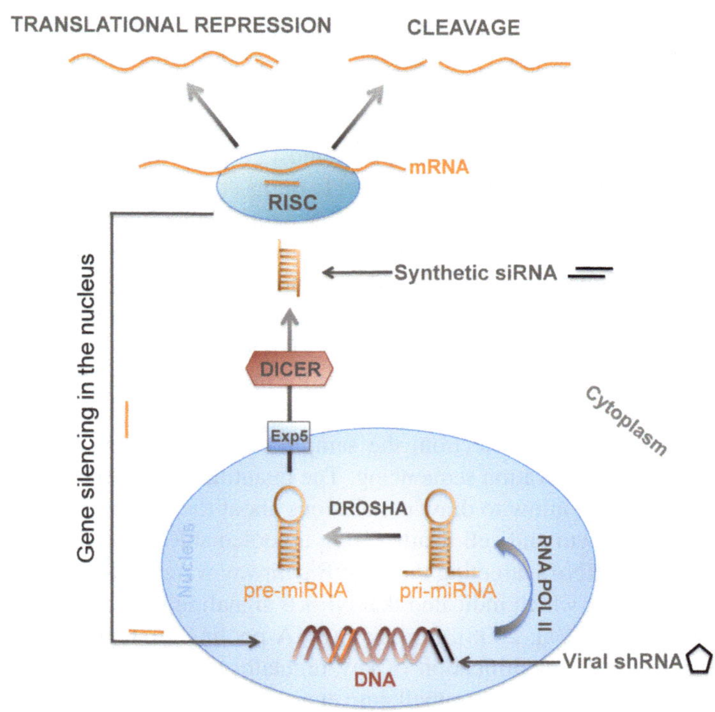

Fig. 9.1 Schematic representation of RNA interference (RNAi). Endogenously encoded microRNAs or virally integrated shRNA are transcribed from the genome by RNA polymerase II into pri-miRNAs. These pri-miRNAs are further processed by the RNAse III enzyme Drosha. The resulting pre-miRNA is then transported from the nucleus into the cytoplasm via Exportin 5 (Exp5) where it is processed by Dicer into a 19–25 bp-long microRNA duplex structure. The mature miRNA is incoporated into the RNA-induced silencing complex (RISC) where it is unwound and preferentially paired to the 3′ untranslated region (UTR) of mRNA. Perfect base-pairing leads to cleavage and degradation of complement mRNA while imperfect base-pairing will result in translational repression. Long dsRNAs are cleaved by Dicer into small interfering RNAs (siRNAs) which are also incoporated into RISC causing downregulation of target gene expression. siRNAs can also be synthetically introduced to the cell to trigger RNAi-mediated gene silencing. siRNAs complementary to promoter regions have been shown to cause transcriptional gene silencing in the nucleus, most likely by heterochromatin modification via histone methylases. However, the exact molecular mechanism of this phenomenon is currently not fully understood

decrease the susceptibility of the siRNA duplex to nuclease attacks (Behlke 2008; Monia et al. 1996; Pandolfi et al. 1999).

siRNAs can also be produced inside the cell by shRNA-based expression vectors. The first expression vector encoding an shRNA hairpin, that is transcribed from an RNA polymerase III promotor, was published in 2002 by Brummelkamp et al. (Brummelkamp et al. 2002). The authors showed that expression of siRNAs using this vector caused stable and effective down-regulation of target gene expression. Furthermore, continuous expression of siRNAs enabled the analysis of loss-of-function phenotypes that develop over longer time periods. Synthetic siRNAs and shRNA-expressing plasmids can also be reverse transfected using multiwell plates or cell-based microarrays (Ziauddin and Sabatini 2001). In primary cells that are difficult to transfect with plasmids or synthetic siRNAs, virally based shRNA delivery systems are often used. To date, several shRNA libraries in adeno-, retro-, or lentiviral vector backbones in an almost genome-wide scale are available (Arts et al. 2003; Bayona-Bafaluy et al. 2011; Berns et al. 2004; Moffat et al. 2006; Paddison et al. 2004a; Silva et al. 2005). Stable expression of the respective shRNA is achieved by integration of the virus into the host genome. In addition to classical shRNA libraries, pooled shRNA libraries comprising one shRNA and a specific barcode per vector have recently become available (Paddison et al. 2004a). These libraries can be used to infect a batch of cells to quickly identify genes required for proliferation and cell survival. The virus titer for these pooled shRNA libraries has to be adjusted so that the vector integration per target cells is not more than one. After virus infection and selection, the sample can be analyzed by microarrays techniques or next-generation sequencing. The quantitative presence or absence of certain barcodes will allow to draw conclusions about the effects of the respective shRNA on cell growth and cell death. Using a pooled shRNA strategy, Ngo et al. could show that shRNAs targeting the NF-κB pathway were depleted in a subgroup of B-cell lymphoma which indicated that NF-κB signaling was crucial for survival of these cells (Ngo et al. 2006). Similar shRNA pooling strategies on a genome-wide scale led to the identification of key regulators of cancer cell survival and targets of small molecules (Brummelkamp et al. 2006; Schlabach et al. 2008; Silva et al. 2008).

Many scientists use inducible RNAi systems to control siRNA expression in a time and dose-specific manner. Most of the inducible vehicles are Tet or Cre/lox P-based systems available in viral and non-viral vectors. Both systems can be used in cell culture and transgenic animals. Tetracycline-inducible systems allow reversible regulation of shRNA expression (on/off switch). However, Tet-regulated promotors might be leaky and potentially cause stimulation-independent down-regulation of target genes to a certain extent. In contrast, site-specific DNA recombination via lox P sites by Cre is irreversible and enables stringent shRNA expression. In transgenic animals, the Cre/lox P system is a common technique to induce site-specific recombination at a defined locus. Several genes of interest could be efficiently downregulated in mouse embryonic stem cells (ESCs) using a retroviral vector and loxP site-directed recombination (Wang 2010). In this study, the vector contained a microRNA-embedded short-hairpin RNA (shRNAmir) to trigger the RNAi mechanism. The idea to express shRNAs in a similar way to primary miRNA transcripts came from the labs of Greg Hannon (CSHL) and Steve Elledge (Harvard) (Paddison et al. 2004b). They showed that shRNAmirs trigger the RNAi pathway through a more natural route leading to specific and effective gene silencing. To date, a lentiviral shRNAmir library covering around 18,000 genes is commercially available. The vectors of this library named pGIPZ and pTRIPZ contain a fluorescence marker and a molecular barcode enabling easy identification of transduced cells and multiplex screening in pools (Chang et al. 2006).

In the past few years, the design of RNAi reagents has constantly improved. New software tools that predict the efficiency of RNAi reagents and their potential off-target effects have become available. One of these tools is NEXT-RNAi that allows prediction and evaluation of siRNA and long dsRNA for all annotated genomes (Horn et al. 2010). However, the complete elimination of off-target effects remains a challenge. Off-target effect means that the applied RNAi reagent does not only target the gene of interest, but also other transcripts leading to an increased rate of false-positive or false-negative candidate genes in an RNAi screen (Jackson et al. 2003). It has been shown that off-targeted genes often contain sequence homologies between the 3′ UTR of the transcript and the seed region of the siRNA antisense strand comprising six nucleotides in positions 2–7 (Birmingham et al. 2006; Lin et al. 2005). The observed unintended silencing is likely to occur through a mechanism similar to miRNA-mediated gene regulation. Thus, improved siRNA design algorithms and appropriate filters can significantly reduce off-target effects. It has also been observed that unintended actions of RNAi reagents are concentration-dependent (Semizarov et al. 2003). A low concentration of an siRNA can significantly reduce its off-target effects, but may also lead to insufficient target gene knockdown. Pooling strategies are often used to circumvent this problem. siRNA pools usually contain 3–4 individual siRNAs that target the same transcript at different sites. Therefore, the concentration of each single RNAi reagent is reduced while the functionality of the pool remains high. A similar principle of operation is utilized by endoribonuclease-prepared siRNAs (esiRNA). EsiRNAs are pools of short overlapping fragments of siRNAs that can be generated by *in vitro* cleavage of long dsRNA with *Escherichia coli* RNase III. As a result, the mixture of these multiple short RNA fragments that all target the same gene transcript can be used at a very low concentration to cause effective gene silencing. A further reduction of off-target effects can be achieved by chemical modification of siRNAs such as 2′-*o*-methyl ribosyl substitution in the guide strand (Jackson et al. 2006).

9.2.2 High-Throughput RNAi Screens for Apoptosis Modulators

For the successful outcome of a high-throughput screen, the setup of an appropriate readout assay is as important as the choice of the RNAi reagent. Cell viability can quickly be determined by quantification of ATP levels, membrane integrity or cellular redox potential (Crouch et al. 1993; Vistica et al. 1991). Cell viability assays are often used to identify modulators of cell death despite the indirect nature of the measurement. An observed reduction in cell viability can be a result of necrosis, apoptosis, other forms of cell death or caused by decreased cellular proliferation. To address whether a gene is involved in regulation of apoptosis caspase activation can be monitored using fluorescent or luminescent-based substrates (Antczak et al. 2009). Another possibility is to quantify AnnexinV/PI positive cells (Vermes et al. 1995) or fragmented DNA (Nicoletti et al. 1991) by high-throughput flow-cytometric analysis (FACS). However, these assays have some disadvantages compared to cell viability readouts such as higher complexity, consumed time, and reagent costs. Furthermore, right timing of the measurements especially when determining caspase activity may be crucial for the outcome of the screen. Therefore, most high-throughput RNAi screens for cell death modulators are based on cell viability measurements in the initial screen setup followed by further candidate evaluation in more specific assays.

Today, over 500 studies performing high-throughput RNAi screens have been reported. Many of these studies have significantly contributed to our understanding of cell death and cell survival pathways.

The first large-scale RNAi screens for modulators of cell viability were performed in *C. elegans* (Kamath et al. 2003) and cultured *Drosophila* cells (Boutros et al. 2004). In the *C. elegans* screen 1,170 out of 16,757 dsRNA that were encoded in *E. coli* strains and fed to the worms resulted in lethality or sterility. In the *Drosophila* screen, the authors identified 438 out of 19,470 dsRNAs that were essential for cell growth and cell viability. Among those genes prominent cell death modulators with human homologues such as cIAP1 could be found. Another genome-wide RNAi screen in *Drosophila* cells identified several genes implicated in the regulation of caspase activation that had previously not been linked to apoptosis signaling such as the *N*-acetyltransferase ARD1 or the candidate tumor suppressor Charlatan (Yi et al. 2007). *Drosophila* proved to be a good model system for the identification of conserved genes that were missed in mammalian screens due to functional redundancy. Nowadays, large-scale RNAi screens cannot only be conduced in cultured *Drosophila* cells, but also in transgenic flies using the Vienna *Drosophila* RNAi collection (VDRC) based on the UAS-Gal4 system (Dietzl et al. 2007).

The first RNAi screen for apoptosis modulators in human cells was carried out by Aza-Blanc et al. (Aza-Blanc et al. 2003). The authors used the death ligand TRAIL (TNF-related apoptosis-inducing ligand) to induce apoptosis in the cervix carcinoma cell line HeLa. They screened a siRNA library comprising 510 genes including known and predicted kinases. Using this approach the previously undescribed proteins DOBI (downstream of BID) and MIRSA (Mina53-related suppressor of apoptosis) could be identified as modulators of TRAIL-induced apoptosis.

In the past few years, large-scale RNAi approaches conducted *in vivo* have led to the discovery of novel gene functions with implications in cancer cell survival and lethality. Using a pri-miRNA-based shRNA library in a mouse model of hepatocellular carcinoma several novel tumor suppressor genes could be identified (Dickins et al. 2005; Zender et al. 2008). Among the described genes, tumor suppressors implicated in the control of cell proliferation and apoptosis could be found.

Michael Green and colleagues used a mouse shRNA library to screen for factors that reduce transcription of ATF5, a gene that is highly expressed in malignant glioma (Sheng et al. 2010). The authors were able to reconstruct a signaling cascade where RAS/MAPK signaling caused induction of the transcription factor CREB3L2 that activated expression of ATF5 that in turn led to transcription of the antiapoptotic protein Mcl1 thereby preventing cell death. The knowledge of this signaling network has enabled the use of small molecules that interfere with this pathway such as the RAF kinase blocker sorafenib as treatment option for brain cancer.

9.3 Synthetic Lethality

Cell death caused by simultaneous perturbation of two genes is referred to as synthetic lethality (see Fig. 9.2). The concept of synthetic lethal genes was first described in a classical forward genetic approach in *Drosophila pseudobscura* (Dobzhansky et al. 1965). It was observed that lethal chromosomes could emerge by genetic recombination between nonlethal chromosomes. This lethal cross-over was referred to as "synthetic lethal."

The first system-wide analysis of synthetic lethality was performed in the yeast *Saccharomyces cerevisiae* (*S. cerevisiae*). Knockout mutants of each of the

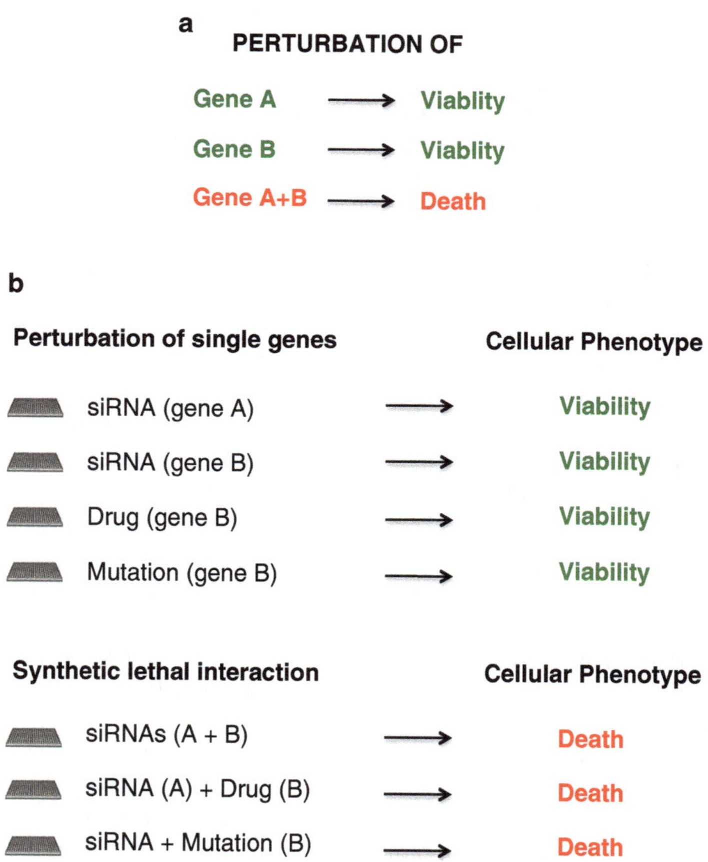

Fig. 9.2 Concept of synthetic lethality. (**a**) A genetic interaction is present when the phenotype of combined perturbation of gene A and B is deviating from the phenotype that is caused by perturbation of gene A or gene B alone. If the consequence of combined perturbation is cell death, the synergy is called synthetic lethal interaction. (**b**) Synthetic lethality can be achieved by knockdown of two target genes, knockdown of one target gene and drug-induced perturbation of the second gene or knockdown of one target gene in a mutant background

approximately 6,000 genes of *S. cerevisiae* have been created and analyzed for viability and growth potential. Interestingly, only ~1,100 genes were required for growth on rich glucose medium indicating that essential signaling pathways in yeast are highly redundant (Giaever et al. 2002). Synthetic lethal screens where two genes were perturbed at the same time revealed that genes with common synthetic lethal partners have an increased probability of belonging to the same signaling pathway, often acting together in the same protein complex (Dixon et al. 2008; Tong et al. 2004; Ye et al. 2005).

Nowadays, synthetic lethal approaches are frequently applied in the field of cancer research. Screens are often designed to unravel genetic interactions with oncogenes or tumor suppressor genes such as RAS and p53. The results of these research attempts helped to reconstruct signaling networks and determine intersection points that are required for cell death or cell survival. They also help to explain why some cancer cells respond to certain drugs while others are resistant.

A good example of a genetic synthetic interaction that is already exploited in cancer therapy is the BRCA-PARP synthetic lethal interaction. BRCA1 and BRCA2 mutations are frequently found in breast, ovarian, colon, and prostate cancer. Loss-of-function of BRCA1 or BRCA2 interferes with proper DNA damage repair, thus increasing the chance of a cell to accumulate mutations and become malignant. Using a synthetic lethal approach, PARP1 was identified to genetically interact with BRCA1 and BRCA2 (Bryant et al. 2005; Farmer et al. 2005; Lord and Ashworth 2008). Under normal circumstances, DNA is repaired through homologous recombination by BRCA1 and BRCA2 or base-excision repair by PARP1. Cells that have lost BRCA1 or BRCA2 can still survive because other DNA-repair mechanisms can compensate for the loss of homologous recombination. However, loss of both proteins, PARP1 and BRCA1 or BRCA2, abolished DNA repair and rapidly induced apoptosis. Fortunately, BRCA1 and BRCA2 mutant cells are around 1,000-fold more sensitive to PARP inhibition than normal cells with intact DNA repair pathways. As a consequence, PARP inhibitors can be used to induce cancer-specific apoptosis with a large therapeutic window. Based on these studies, the PARP inhibitors olaparib (Astra Zeneca) and iniparib (Sanofi Aventis) have been developed and are currently evaluated in phase II clinical trials. This very rapid translation of research results into clinical applications has generated a lot of excitement for synthetic lethal approaches in tumor therapy.

The first large-scale synthetic lethal screening approach in human cells was performed in a RAS mutant background using genome-wide RNAi. The oncogene RAS is frequently mutated in a variety of cancers and causes the continuous transmission of cell survival and proliferation signals independent of extracellular signals (Bos 1989). As a result, RAS-mutant cancer cells proliferate in an uncontrolled manner and are often insensitive to apoptotic stimuli. Many scientists have tried to pharmacologically inactivate mutant RAS, which however have not been successful. At the moment, alternative pathway inhibitors such as transfarnesylthiosalicylic acid (FTS), which impairs translocation of RAS to the plasma membrane, are evaluated as drug candidates (Rotblat et al. 2008). In an attempt to identify essential components and potential novel drug targets for the survival of RAS mutant cells, RNAi screens were performed in 19 different RAS mutant and wild-type cells. The utilized shRNA-based library comprised around 1,000 genes including kinases, phosphatases, and other cancer-related genes. Using this targeted approach, the noncanonical IkappaB kinase TBK1 as well as the serine/threonine kinase STK33 could be identified as synthetic lethal interaction partners of oncogenic KRAS (Barbie et al. 2009; Scholl et al. 2009). Genome-wide high throughput screen for synthetic lethal RAS interactors were also conducted in isogenic DLD1 and HCT116 colon cancer cells that either carry an endogenous activating KRAS mutation or a wild-type allele (Luo et al. 2009; Wang et al. 2010a). By integration of loss-of-function and gene expression data, a previously underestimated role of Ras in mitotic progression involving the mitotic checkpoint kinase PLK1 (Luo et al. 2009). In addition, SNAIL2 was found to be an essential factor for the survival of KRAS mutant cells that have undergone epithelial–mesenchymal transition (EMT) (Wang et al. 2010a).

Besides RAS mutations, deletions of the adenomatous polyposis coli protein (APC) occur in more than 80% of all colon cancers leading to accumulation of nuclear β-catenin and uncontrolled proliferation. Zhang et al. (2010) recently showed that a combination of TRAIL and RAc (all-trans-retinyl acetate) specifically kills APC-deficient premalignant tumor cells. This synergistic interaction was caused by RAc-mediated upregulation of TRAIL receptors and concomitant repression of cFLIP through activation of c-myc by b-catenin and could in future be exploited in the prevention of colorectal cancer.

The tumor suppressor p53 that is mutated in more than 50% of all human cancers (Hollstein et al. 1991) is another desired interaction candidate for synthetic lethal screening approaches. In normal cells, p53 is a guardian of the DNA repair machinery (Dasika et al. 1999). Upon DNA damage, induction of p53 ensures that the cell cycle is arrested at the G1 stage allowing for proper DNA repair. In case the damage is too severe to be repaired, p53 quickly triggers a cell suicide program, reducing the chance that a cell acquires decisive mutations and becomes cancerous. Several genes that are involved in DNA damage repair such as ATM and Chk1 as well as several kinases have been identified as synthetic lethal interaction partners of p53 (Baldwin et al. 2010; Jiang et al. 2009; Wang et al. 2004, 2010b).

In summary, synthetic lethal screens greatly improved our understanding of cell survival and cell death pathways in cancer. Furthermore, they provided insights into oncogenic network addiction and helped to develop targeted therapies for cancer treatment.

9.4 Outlook

To dissect the genetic circuitry that underlie life and death decisions of cells, advances in technology and experimental paradigms promise to provide new mechanistic and physiological insights. Future systematic loss-of-function approaches will involve more complex assays that give a broader view of a phenotype. For example, it will be possible to exactly determine whether a cell has undergone apoptosis, necrosis or other forms of cell death. Combination of morphological signatures will allow a system-wide analysis and clustering of phenotypes based on their features. Furthermore, time-resolved dissection of phenotypes on a single cell bases will allow to determine cell to cell variability and cell type specificity while phenotypic similarities and genetic interaction profiles enable the association of genes with biological processes (Fuchs et al. 2010; Horn et al. 2011).

However, there are also challenges that have to be addressed in the future such as standardized annotation formats for system-wide loss-of-function experiments, availability of instruments for multiparametric data acquisition, computing power or long-term storage of large data sets.

In summary, systematic loss-of-function approaches significantly contributed to our understanding of apoptosis and cell survival networks. The combination of novel tools and well-established genetic approaches such as synthetic lethality will provide medical relevant models on the regulation of apoptosis in normal and cancer cells.

Acknowledgements We acknowledge funding from the EraSysBio + ApoNET Network and the Helmholtz Alliance for Systems biology.

References

Abrams JM, White K, Fessler LI, Steller H (1993) Programmed cell death during Drosophila embryogenesis. Development 117:29–43

Antczak C, Takagi T, Ramirez CN, Radu C, Djaballah H (2009) Live-cell imaging of caspase activation for high-content screening. J Biomol Screen 14:956–969

Arts GJ, Langemeijer E, Tissingh R, Ma L, Pavliska H, Dokic K, Dooijes R, Mesic E, Clasen R, Michiels F et al (2003) Adenoviral vectors expressing siRNAs for discovery and validation of gene function. Genome Res 13:2325–2332

Aza-Blanc P, Cooper CL, Wagner K, Batalov S, Deveraux QL, Cooke MP (2003) Identification of modulators of TRAIL-induced apoptosis via RNAi-based phenotypic screening. Mol Cell 12:627–637

Baldwin A, Grueneberg DA, Hellner K, Sawyer J, Grace M, Li W, Harlow E, Munger K (2010) Kinase requirements in human cells: V. Synthetic lethal interactions between p53 and the protein kinases SGK2 and PAK3. Proc Natl Acad Sci USA 107:12463–12468

Barbie DA, Tamayo P, Boehm JS, Kim SY, Moody SE, Dunn IF, Schinzel AC, Sandy P, Meylan E, Scholl C et al (2009) Systematic RNA interference reveals that oncogenic KRAS-driven cancers require TBK1. Nature 462:108–112

Bayona-Bafaluy MP, Sanchez-Cabo F, Fernandez-Silva P, Perez-Martos A, Enriquez JA (2011) A genome-wide shRNA screen for new OxPhos related genes. Mitochondrion 11(3):467–475

Behlke MA (2008) Chemical modification of siRNAs for in vivo use. Oligonucleotides 18:305–319

Bentwich I, Avniel A, Karov Y, Aharonov R, Gilad S, Barad O, Barzilai A, Einat P, Einav U, Meiri E et al (2005) Identification of hundreds of conserved and nonconserved human microRNAs. Nat Genet 37:766–770

Berns K, Hijmans EM, Mullenders J, Brummelkamp TR, Velds A, Heimerikx M, Kerkhoven RM, Madiredjo M, Nijkamp W, Weigelt B et al (2004) A large-scale RNAi screen in human cells identifies new components of the p53 pathway. Nature 428:431–437

Birmingham A, Anderson EM, Reynolds A, Ilsley-Tyree D, Leake D, Fedorov Y, Baskerville S, Maksimova E, Robinson K, Karpilow J et al (2006) 3′ UTR seed matches, but not overall identity, are associated with RNAi off-targets. Nat Methods 3:199–204

Bos JL (1989) ras oncogenes in human cancer: a review. Cancer Res 49:4682–4689

Boutros M, Ahringer J (2008) The art and design of genetic screens: RNA interference. Nat Rev Genet 9:554–566

Boutros M, Kiger AA, Armknecht S, Kerr K, Hild M, Koch B, Haas SA, Paro R, Perrimon N (2004) Genome-wide RNAi analysis of growth and viability in Drosophila cells. Science 303:832–835

Brenner S (1974) The genetics of Caenorhabditis elegans. Genetics 77:71–94

Brummelkamp TR, Bernards R, Agami R (2002) A system for stable expression of short interfering RNAs in mammalian cells. Science 296:550–553

Brummelkamp TR, Fabius AW, Mullenders J, Madiredjo M, Velds A, Kerkhoven RM, Bernards R, Beijersbergen RL (2006) An shRNA barcode screen provides insight into cancer cell vulnerability to MDM2 inhibitors. Nat Chem Biol 2:202–206

Bryant HE, Schultz N, Thomas HD, Parker KM, Flower D, Lopez E, Kyle S, Meuth M, Curtin NJ, Helleday T (2005) Specific killing of BRCA2-deficient tumours with inhibitors of poly(ADP-ribose) polymerase. Nature 434:913–917

Cecconi F (1999) Apaf1 and the apoptotic machinery. Cell Death Differ 6:1087–1098

Chai J, Du C, Wu JW, Kyin S, Wang X, Shi Y (2000) Structural and biochemical basis of apoptotic activation by Smac/DIABLO. Nature 406:855–862

Chang K, Elledge SJ, Hannon GJ (2006) Lessons from Nature: microRNA-based shRNA libraries. Nat Methods 3:707–714

Crouch SP, Kozlowski R, Slater KJ, Fletcher J (1993) The use of ATP bioluminescence as a measure of cell proliferation and cytotoxicity. J Immunol Methods 160:81–88

Dasika GK, Lin SC, Zhao S, Sung P, Tomkinson A, Lee EY (1999) DNA damage-induced cell cycle checkpoints and DNA strand break repair in development and tumorigenesis. Oncogene 18:7883–7899

Dickins RA, Hemann MT, Zilfou JT, Simpson DR, Ibarra I, Hannon GJ, Lowe SW (2005) Probing tumor phenotypes using stable and regulated synthetic microRNA precursors. Nat Genet 37:1289–1295

Dietzl G, Chen D, Schnorrer F, Su KC, Barinova Y, Fellner M, Gasser B, Kinsey K, Oppel S, Scheiblauer S et al (2007) A genome-wide transgenic RNAi library for conditional gene inactivation in Drosophila. Nature 448:151–156

Dixon SJ, Fedyshyn Y, Koh JL, Prasad TS, Chahwan C, Chua G, Toufighi K, Baryshnikova A, Hayles J, Hoe KL et al (2008) Significant conservation of synthetic lethal genetic interaction networks between distantly related eukaryotes. Proc Natl Acad Sci USA 105:16653–16658

Dobzhansky T, Spassky B, Anderson W (1965) Bichromosomal synthetic semilethals in Drosophila pseudoobscura. Proc Natl Acad Sci USA 53:482–486

Elbashir SM, Harborth J, Weber K, Tuschl T (2002) Analysis of gene function in somatic mammalian cells using small interfering RNAs. Methods 26:199–213

Ellis HM, Horvitz HR (1986) Genetic control of programmed cell death in the nematode C. elegans. Cell 44:817–829

Ellis RE, Jacobson DM, Horvitz HR (1991) Genes required for the engulfment of cell corpses during programmed cell death in Caenorhabditis elegans. Genetics 129:79–94

Epstein HF, Waterston RH, Brenner S (1974) A mutant affecting the heavy chain of myosin in Caenorhabditis elegans. J Mol Biol 90:291–300

Farmer H, McCabe N, Lord CJ, Tutt AN, Johnson DA, Richardson TB, Santarosa M, Dillon KJ, Hickson I, Knights C et al (2005) Targeting the DNA repair defect in BRCA mutant cells as a therapeutic strategy. Nature 434:917–921

Ferguson EL, Sternberg PW, Horvitz HR (1987) A genetic pathway for the specification of the vulval cell lineages of Caenorhabditis elegans. Nature 326:259–267

Fire A, Xu S, Montgomery MK, Kostas SA, Driver SE, Mello CC (1998) Potent and specific genetic interference by double-stranded RNA in Caenorhabditis elegans. Nature 391:806–811

Fuchs F, Pau G, Kranz D, Sklyar O, Budjan C, Steinbrink S, Horn T, Pedal A, Huber W, Boutros M (2010) Clustering phenotype populations by genome-wide RNAi and multiparametric imaging. Mol Syst Biol 6:370

Giaever G, Chu AM, Ni L, Connelly C, Riles L, Veronneau S, Dow S, Lucau-Danila A, Anderson K, Andre B et al (2002) Functional profiling of the Saccharomyces cerevisiae genome. Nature 418:387–391

Good L (2003) Translation repression by antisense sequences. Cell Mol Life Sci 60:854–861

Han J, Lee Y, Yeom KH, Kim YK, Jin H, Kim VN (2004) The Drosha-DGCR8 complex in primary microRNA processing. Genes Dev 18:3016–3027

Hay BA, Wassarman DA, Rubin GM (1995) Drosophila homologs of baculovirus inhibitor of apoptosis proteins function to block cell death. Cell 83:1253–1262

Hengartner MO, Horvitz HR (1994) C. elegans cell survival gene ced-9 encodes a functional homolog of the mammalian proto-oncogene bcl-2. Cell 76:665–676

Hengartner MO, Ellis RE, Horvitz HR (1992) Caenorhabditis elegans gene ced-9 protects cells from programmed cell death. Nature 356:494–499

Hollstein M, Sidransky D, Vogelstein B, Harris CC (1991) p53 mutations in human cancers. Science 253:49–53

Horn T, Sandmann T, Boutros M (2010) Design and evaluation of genome-wide libraries for RNA interference screens. Genome Biol 11:R61

Horn T, Sandmann T, Fischer B, Axelsson E, Huber W, Boutros M (2011) Mapping of signaling networks through synthetic genetic interaction analysis by RNAi. Nat Methods 8:341–346

Horvitz HR, Shaham S, Hengartner MO (1994) The genetics of programmed cell death in the nematode Caenorhabditis elegans. Cold Spring Harb Symp Quant Biol 59:377–385

Hu M, Wang M, Zhu H, Zhang L, Zhang H, Sun L (2010) Preparation and structures of enantiomeric dinuclear zirconium and hafnium complexes containing two homochiral N atoms, and their catalytic property for polymerization of rac-lactide. Dalton Trans 39:4440–4446

Huang DC, Strasser A (2000) BH3-Only proteins-essential initiators of apoptotic cell death. Cell 103:839–842

Igaki T, Yamamoto-Goto Y, Tokushige N, Kanda H, Miura M (2002) Down-regulation of DIAP1 triggers a novel Drosophila cell death pathway mediated by Dark and DRONC. J Biol Chem 277:23103–23106

Jackson AL, Bartz SR, Schelter J, Kobayashi SV, Burchard J, Mao M, Li B, Cavet G, Linsley PS (2003) Expression profiling reveals off-target gene regulation by RNAi. Nat Biotechnol 21:635–637

Jackson AL, Burchard J, Leake D, Reynolds A, Schelter J, Guo J, Johnson JM, Lim L, Karpilow J, Nichols K et al (2006) Position-specific chemical modification of siRNAs reduces "off-target" transcript silencing. RNA 12:1197–1205

Jiang H, Reinhardt HC, Bartkova J, Tommiska J, Blomqvist C, Nevanlinna H, Bartek J, Yaffe MB, Hemann MT (2009) The combined status of ATM and p53 link tumor development with therapeutic response. Genes Dev 23:1895–1909

Kamath RS, Fraser AG, Dong Y, Poulin G, Durbin R, Gotta M, Kanapin A, Le Bot N, Moreno S, Sohrmann M et al (2003) Systematic functional analysis of the Caenorhabditis elegans genome using RNAi. Nature 421:231–237

Karlas A, Machuy N, Shin Y, Pleissner KP, Artarini A, Heuer D, Becker D, Khalil H, Ogilvie LA, Hess S et al (2010) Genome-wide RNAi screen identifies human host factors crucial for influenza virus replication. Nature 463:818–822

Kumar S, Doumanis J (2000) The fly caspases. Cell Death Differ 7:1039–1044

Lau NC, Lim LP, Weinstein EG, Bartel DP (2001) An abundant class of tiny RNAs with probable regulatory roles in Caenorhabditis elegans. Science 294:858–862

Lin X, Ruan X, Anderson MG, McDowell JA, Kroeger PE, Fesik SW, Shen Y (2005) siRNA-mediated off-target gene silencing triggered by a 7 nt complementation. Nucleic Acids Res 33:4527–4535

Lisi S, Mazzon I, White K (2000) Diverse domains of THREAD/DIAP1 are required to inhibit apoptosis induced by REAPER and HID in Drosophila. Genetics 154:669–678

Liu Z, Sun C, Olejniczak ET, Meadows RP, Betz SF, Oost T, Herrmann J, Wu JC, Fesik SW (2000) Structural basis for binding of Smac/DIABLO to the XIAP BIR3 domain. Nature 408:1004–1008

Lockshin RA, Williams CM (1965) Programmed cell death–I cytology of degeneration in the intersegmental muscles of the Pernyi Silkmoth. J Insect Physiol 11:123–133

Lord CJ, Ashworth A (2008) Targeted therapy for cancer using PARP inhibitors. Curr Opin Pharmacol 8:363–369

Luo J, Emanuele MJ, Li D, Creighton CJ, Schlabach MR, Westbrook TF, Wong KK, Elledge SJ (2009) A genome-wide RNAi screen identifies multiple synthetic lethal interactions with the Ras oncogene. Cell 137:835–848

Martins LM, Iaccarino I, Tenev T, Gschmeissner S, Totty NF, Lemoine NR, Savopoulos J, Gray CW, Creasy CL, Dingwall C et al (2002) The serine protease Omi/HtrA2 regulates apoptosis by binding XIAP through a reaper-like motif. J Biol Chem 277:439–444

Meister G, Landthaler M, Patkaniowska A, Dorsett Y, Teng G, Tuschl T (2004) Human Argonaute2 mediates RNA cleavage targeted by miRNAs and siRNAs. Mol Cell 15:185–197

Mette MF, Aufsatz W, van der Winden J, Matzke MA, Matzke AJ (2000) Transcriptional silencing and promoter methylation triggered by double-stranded RNA. EMBO J 19:5194–5201

Moffat J, Grueneberg DA, Yang X, Kim SY, Kloepfer AM, Hinkle G, Piqani B, Eisenhaure TM, Luo B, Grenier JK et al (2006) A lentiviral RNAi library for human and mouse genes applied to an arrayed viral high-content screen. Cell 124:1283–1298

Monia BP, Johnston JF, Sasmor H, Cummins LL (1996) Nuclease resistance and antisense activity of modified oligonucleotides targeted to Ha-ras. J Biol Chem 271:14533–14540

Neely GG, Kuba K, Cammarato A, Isobe K, Amann S, Zhang L, Murata M, Elmen L, Gupta V, Arora S et al (2010) A global in vivo Drosophila RNAi screen identifies NOT3 as a conserved regulator of heart function. Cell 141:142–153

Ngo VN, Davis RE, Lamy L, Yu X, Zhao H, Lenz G, Lam LT, Dave S, Yang L, Powell J et al (2006) A loss-of-function RNA interference screen for molecular targets in cancer. Nature 441:106–110

Nicoletti I, Migliorati G, Pagliacci MC, Grignani F, Riccardi C (1991) A rapid and simple method for measuring thymocyte apoptosis by propidium iodide staining and flow cytometry. J Immunol Methods 139:271–279

Paddison PJ, Cleary M, Silva JM, Chang K, Sheth N, Sachidanandam R, Hannon GJ (2004a) Cloning of short hairpin RNAs for gene knockdown in mammalian cells. Nat Methods 1:163–167

Paddison PJ, Silva JM, Conklin DS, Schlabach M, Li M, Aruleba S, Balija V, O'Shaughnessy A, Gnoj L, Scobie K et al (2004b) A resource for large-scale RNA-interference-based screens in mammals. Nature 428:427–431

Pandolfi D, Rauzi F, Capobianco ML (1999) Evaluation of different types of end-capping modifications on the stability of oligonucleotides toward 3′- and 5′-exonucleases. Nucleosides Nucleotides 18:2051–2069

Patton EE, Zon LI (2001) The art and design of genetic screens: zebrafish. Nat Rev Genet 2:956–966

Preall JB, Sontheimer EJ (2005) RNAi: RISC gets loaded. Cell 123:543–545

Rodriguez A, Chen P, Oliver H, Abrams JM (2002) Unrestrained caspase-dependent cell death caused by loss of Diap1 function requires the Drosophila Apaf-1 homolog, Dark. EMBO J 21:2189–2197

Rotblat B, Ehrlich M, Haklai R, Kloog Y (2008) The Ras inhibitor farnesylthiosalicylic acid (Salirasib) disrupts the spatiotemporal localization of active Ras: a potential treatment for cancer. Methods Enzymol 439:467–489

Salvesen GS, Duckett CS (2002) IAP proteins: blocking the road to death's door. Nat Rev Mol Cell Biol 3:401–410

Schlabach MR, Luo J, Solimini NL, Hu G, Xu Q, Li MZ, Zhao Z, Smogorzewska A, Sowa ME, Ang XL et al (2008) Cancer proliferation gene discovery through functional genomics. Science 319:620–624

Scholl C, Frohling S, Dunn IF, Schinzel AC, Barbie DA, Kim SY, Silver SJ, Tamayo P, Wadlow RC, Ramaswamy S et al (2009) Synthetic lethal interaction between oncogenic KRAS dependency and STK33 suppression in human cancer cells. Cell 137:821–834

Schwartz LM, Osborne BA (1994) Ced-3/ICE: evolutionarily conserved regulation of cell death. Bioessays 16:387–389

Semizarov D, Frost L, Sarthy A, Kroeger P, Halbert DN, Fesik SW (2003) Specificity of short interfering RNA determined through gene expression signatures. Proc Natl Acad Sci USA 100:6347–6352

Sheng Z, Li L, Zhu LJ, Smith TW, Demers A, Ross AH, Moser RP, Green MR (2010) A genome-wide RNA interference screen reveals an essential CREB3L2-ATF5-MCL1 survival pathway in malignant glioma with therapeutic implications. Nat Med 16:671–677

Silva JM, Li MZ, Chang K, Ge W, Golding MC, Rickles RJ, Siolas D, Hu G, Paddison PJ, Schlabach MR et al (2005) Second-generation shRNA libraries covering the mouse and human genomes. Nat Genet 37:1281–1288

Silva JM, Marran K, Parker JS, Silva J, Golding M, Schlabach MR, Elledge SJ, Hannon GJ, Chang K (2008) Profiling essential genes in human mammary cells by multiplex RNAi screening. Science 319:617–620

Srinivasula SM, Datta P, Fan XJ, Fernandes-Alnemri T, Huang Z, Alnemri ES (2000) Molecular determinants of the caspase-promoting activity of Smac/DIABLO and its role in the death receptor pathway. J Biol Chem 275:36152–36157

St Johnston D (2002) The art and design of genetic screens: Drosophila melanogaster. Nat Rev Genet 3:176–188

Sulston JE, Horvitz HR (1977) Post-embryonic cell lineages of the nematode, Caenorhabditis elegans. Dev Biol 56:110–156

Sulston JE, Horvitz HR (1981) Abnormal cell lineages in mutants of the nematode Caenorhabditis elegans. Dev Biol 82:41–55

Tabara H, Grishok A, Mello CC (1998) RNAi in C. elegans: soaking in the genome sequence. Science 282:430–431

Tong AH, Lesage G, Bader GD, Ding H, Xu H, Xin X, Young J, Berriz GF, Brost RL, Chang M et al (2004) Global mapping of the yeast genetic interaction network. Science 303:808–813

Trent C, Tsuing N, Horvitz HR (1983) Egg-laying defective mutants of the nematode Caenorhabditis elegans. Genetics 104:619–647

Vaux DL, Silke J (2005) IAPs, RINGs and ubiquitylation. Nat Rev Mol Cell Biol 6:287–297

Vermes I, Haanen C, Steffens-Nakken H, Reutelingsperger C (1995) A novel assay for apoptosis. Flow cytometric detection of phosphatidylserine expression on early apoptotic cells using fluorescein labelled Annexin V. J Immunol Methods 184:39–51

Vistica DT, Skehan P, Scudiero D, Monks A, Pittman A, Boyd MR (1991) Tetrazolium-based assays for cellular viability: a critical examination of selected parameters affecting formazan production. Cancer Res 51:2515–2520

Wang J (2010) Efficient gene knockdowns in mouse embryonic stem cells using microRNA-based shRNAs. Methods Mol Biol 650:241–256

Wang M, Sternberg PW (2001) Pattern formation during C. elegans vulval induction. Curr Top Dev Biol 51:189–220

Wang SL, Hawkins CJ, Yoo SJ, Muller HA, Hay BA (1999) The Drosophila caspase inhibitor DIAP1 is essential for cell survival and is negatively regulated by HID. Cell 98:453–463

Wang Y, Decker SJ, Sebolt-Leopold J (2004) Knockdown of Chk1, Wee1 and Myt1 by RNA interference abrogates G2 checkpoint and induces apoptosis. Cancer Biol Ther 3:305–313

Wang Y, Ngo VN, Marani M, Yang Y, Wright G, Staudt LM, Downward J (2010a) Critical role for transcriptional repressor Snail2 in transformation by oncogenic RAS in colorectal carcinoma cells. Oncogene 29:4658–4670

Wang Y, Zhang W, Edelmann L, Kolodner RD, Kucherlapati R, Edelmann W (2010b) Cis lethal genetic interactions attenuate and alter p53 tumorigenesis. Proc Natl Acad Sci USA 107:5511–5515

White K, Grether ME, Abrams JM, Young L, Farrell K, Steller H (1994) Genetic control of programmed cell death in Drosophila. Science 264:677–683

Wu G, Chai J, Suber TL, Wu JW, Du C, Wang X, Shi Y (2000a) Structural basis of IAP recognition by Smac/DIABLO. Nature 408:1008–1012

Wu YC, Stanfield GM, Horvitz HR (2000b) NUC-1, a caenorhabditis elegans DNase II homolog, functions in an intermediate step of DNA degradation during apoptosis. Genes Dev 14:536–548

Ye P, Peyser BD, Spencer FA, Bader JS (2005) Commensurate distances and similar motifs in genetic congruence and protein interaction networks in yeast. BMC Bioinformatics 6:270

Yi CH, Sogah DK, Boyce M, Degterev A, Christofferson DE, Yuan J (2007) A genome-wide RNAi screen reveals multiple regulators of caspase activation. J Cell Biol 179:619–626

Yuan JY, Horvitz HR (1990) The Caenorhabditis elegans genes ced-3 and ced-4 act cell autonomously to cause programmed cell death. Dev Biol 138:33–41

Zender L, Xue W, Zuber J, Semighini CP, Krasnitz A, Ma B, Zender P, Kubicka S, Luk JM, Schirmacher P et al (2008) An oncogenomics-based in vivo RNAi screen identifies tumor suppressors in liver cancer. Cell 135:852–864

Zeng Y, Cullen BR (2002) RNA interference in human cells is restricted to the cytoplasm. RNA 8:855–860

Zhang EE, Liu AC, Hirota T, Miraglia LJ, Welch G, Pongsawakul PY, Liu X, Atwood A, Huss JW 3rd, Janes J et al (2009) A genome-wide RNAi screen for modifiers of the circadian clock in human cells. Cell 139:199–210

Zhang L, Ren X, Alt E, Bai X, Huang S, Xu Z, Lynch PM, Moyer MP, Wen XF, Wu X (2010) Chemoprevention of colorectal cancer by targeting APC-deficient cells for apoptosis. Nature 464:1058–1061

Ziauddin J, Sabatini DM (2001) Microarrays of cells expressing defined cDNAs. Nature 411:107–110

Index

I.N. Lavrik (ed.), *Systems Biology of Apoptosis*, DOI 10.1007/978-1-4614-4009-3,

L

M

O

MIX
Papier aus verantwortungsvollen Quellen
Paper from responsible sources
FSC® C105338

If you have any concerns about our products,
you can contact us on
ProductSafety@springernature.com

In case Publisher is established outside the EU,
the EU authorized representative is:
Springer Nature Customer Service Center GmbH
Europaplatz 3, 69115 Heidelberg, Germany

Printed by Libri Plureos GmbH
in Hamburg, Germany